BIOLOGICAL AND BIOCHEMICAL
APPLICATIONS
OF
ELECTRON SPIN RESONANCE

MONOGRAPHS ON ELECTRON SPIN RESONANCE

Editor: H. M. Assenheim, Israel Atomic Energy Commission

Biological and Biochemical Applications of Electron Spin Resonance

D. J. E. INGRAM

M.A., D.Phil. D.Sc., Hon.D.Sc., F.Inst.P.

Professor of Physics, University of Keele

PLENUM PRESS

NEW YORK

Published in the U.S.A. by
PLENUM PRESS
a division of
PLENUM PUBLISHING CORPORATION
227 West 17th Street, N.Y., N.Y. 10011

Library of Congress Catalog Card Number: 78–86923

First published by
ADAM HILGER LTD
60 ROCHESTER PLACE, LONDON NW1

© D. J. E. INGRAM, 1969

SBN 85274 089 1

Printed in Great Britain by Bell & Bain Ltd, Glasgow

EDITOR'S PREFACE

Ever since E.S.R. spectroscopy began in the early 1950s, Professor Ingram has played an important, an invaluable role in its development and application. Those at all familiar with the subject need no introduction to him, either as scientist or as author. David Ingram was one of the very first to investigate the biological and biochemical applications that have throughout been growing in importance, all the signs seeming to suggest that they may well become the most important applications of all. When inviting him to contribute to this series of monographs, I had no need to marry the man to the subject; the marriage took place nearly a generation ago. There is surely no need for me to excuse the personal character of this preface. To those who know him and work with him, David Ingram is something more than scientist and author.

What these pages portray is the full story of the biological and biochemical applications of E.S.R. up to the present time. They include chapters on such subjects as radiation effects, enzyme interaction, the triplet state and metallo-organic compounds. The monograph will be of value, not only to physicists and chemists working in biomedicine and biochemistry, but also to medical research workers, biologists, and doctors. It will enable them to understand much of what has gone before, and will also, I hope and expect, stimulate further work in a field of vast importance to mankind.

H. M. ASSENHEIM

May 1969

PREFACE

There is no doubt that some of the most important and interesting applications of electron spin resonance are being found, and will continue to be found, in the biochemical and biological fields of study. In common with most other forms of spectroscopy, electron resonance has moved successively from its initial development by the physicists, through an opening up and assessment of various new fields of investigation by the physical and organic chemists, and finally onto the work of the biochemists and physiologists who have applied it to the more complex systems and processes associated with the life sciences.

It would appear, however, that electron spin resonance has moved along this common line of research development somewhat more quickly than most of its predecessors, and this may well be because the biochemists themselves have appreciated its potentialities from the beginning. It is essentially a technique that can detect and characterize the presence of unpaired electrons in passive or active systems, and, at the very time that electron resonance itself was being developed, so the importance of unpaired electrons and free-radical activity was being appreciated by biochemists in general. They were thus quick to realize that an instrument that could detect and monitor unpaired electron concentrations might prove to be an invaluable tool in probing the kinetics of enzyme reactions and similar catalytic processes, as well as being able to probe the structures of metallo-organic molecules, and help elucidate the mechanisms of radiation damage and protection of biological tissue itself. The number of different research groups, working within University Departments of Biochemistry, or Hospital Medical Schools, that have taken up electron resonance studies is indeed a striking confirmation of this, and the present monograph has been written in the certain belief that there are many more such biochemists and medical research workers who would like to initiate such work and may wish to have some introduction to the potentialities and problems of electron resonance as applied in this field.

This book has, therefore, been written primarily for the biochemist, or other scientist interested in the biological or medical fields, who has no specialized knowledge of electronics or microwave physics or any other of the specific techniques associated with electron resonance. The basic principles of the resonance method are therefore first introduced and these are followed by a non-specialist account of the techniques that are employed, but these are presented in such a way that the experimental problems and limitations, as well as the potentialities, can be clearly understood. Thus, although detailed design data are not given in the book itself, references to other texts where they may be found are always included, and it is hoped that this monograph on its own will give sufficient background and confidence for any research worker to initiate his own programme of electron resonance studies. At the same time, it presents an overall view of the way that electron resonance is being applied in the biochemical and biological fields so that it should be of general interest to others, research workers and students alike, who may not be proposing to work in these fields themselves, but would like a better understanding of these new lines of development.

It should possibly be stressed that most of the current lines of investigation are indeed still being developed, and quite new lines will undoubtedly materialize over the next few years. It might, therefore, be argued that this was not the most appropriate time to write such a book, and it would have been better to have waited for a few years until all the lines of development were well established. It would seem, however, that a large number of those in biochemistry and allied fields are now becoming particularly interested in the potentialities of electron resonance, and it would therefore be much more appropriate to try and offer something which may help them to initiate and develop such new lines of investigation, rather than wait until all were established.

It would probably also be fair to say that some of the major lines along which progress will continue to be made in the future are now becoming clear, and the structure of the present book has been drawn up with this in mind. Thus, after the introductory chapter on basic principles, and two chapters summarizing the techniques that are employed, including the spectrometers themselves, the remaining chapters group together current electron resonance studies under the three main headings of present research interest, i.e. (i) Radiation Studies, (ii) Enzyme Studies, (iii) Investigation of metallo-organic compounds, as they occur in proteins and similar substances. The last chapter then summarizes some of the latest techniques and developments and attempts to make some predictions of the way in which future work may progress. Stress is particularly laid here on the power of simultaneous studies with electron resonance and other techniques, such as X-ray crystallography or Mossbauer experiments. The main theme of the book might very well be summarized as an illustration of the way in which electron resonance, in common with other somewhat specialized techniques, is moving from one area of science to another and, in the process, helping to integrate them together into one unified whole.

In writing the book I have become increasingly aware of the growing interest that scientists of different backgrounds now have in probing through the older lines of demarcation, and I should like to express my thanks to many, physicists on the one hand, and biochemists on the other, who have been so willing to discuss their work, and comment on possible future trends. My particular thanks go to Dr E. F. Slade, who has been supervizing various electron resonance studies on biochemical systems in our own laboratories at Keele, and also to my secretary, Mrs Christine Hayston-Reay, who has been so willing to retype the manuscript of the book so many times.

D. J. E. INGRAM

Keele
March 1969

CONTENTS

Chapter 1

Introduction

§1.1

THE POTENTIALITIES OF ELECTRON RESONANCE IN BIOLOGICAL INVESTIGATIONS

Electron spin resonance (E.S.R.) has been finding rapid application in biological and biochemical studies of recent years. These applications arise from its ability to detect unpaired electron concentrations down to a very low level in all kinds of substances, without destroying or modifying the substance in question, and it can also characterize the particular energy states, or locations, of these unpaired electrons. The importance of unpaired electrons in any chemical or biological system is due to the extra energy, and activity, which is normally associated with them, and hence a study of their properties often gives vital clues on the properties or mechanism of the molecular system to which they belong.

Broadly speaking, unpaired electrons can be divided into two groups; in the first are those associated with either the whole molecule or else a large fraction of it. These unpaired electrons move in highly delocalized molecular orbitals and are responsible for the various activities of groups of atoms around the molecule, thus producing the specificity of various free-radical-type reactions. A study of these delocalized unpaired electrons is thus extremely important in any understanding of the mechanism of such processes as irradiation damage of biological tissue, or of the different molecular transient species that are formed in enzyme or other catalytic activity.

The second group of compounds containing unpaired electrons consists of all those in which the unpaired electron is more closely bound to one single atom, rather than embracing many atoms in a delocalized molecular orbital. Such unpaired electrons are normally associated with transition group atoms such as iron, cobalt or nickel, and both their number per atom and their characteristic energies will vary with the particular valency state of the atom with which they are associated. As a result, electron resonance studies of biological and biochemical compounds containing such metallic atoms, or ions, can often give very specific information on the binding and oxidation state of the ion in question. Such valency changes can also be followed continuously during reactions in which enzymes, or other agents, are active, and the simultaneous study of changing free-radical concentrations and changing metallic valency states can be extremely helpful in this connection.

Although the majority of substances containing unpaired electrons can be classified in one of the two main groups outlined above, there are also many other more specific systems which may be difficult to fit into such a broad classification and which deserve mention on their own. There are, for instance, the unpaired electrons associated with semiconductor mechanisms, and considerable attention has been focused on these in recent years, when the possibilities of semiconductor-type energy-level systems have been postulated for proteins and a link between these and possible carcinogenic activity has been suggested. In a similar way considerable investigation by electron resonance techniques has been made of the photosynthetic process, and although the unpaired electrons taking part in this mechanism can hardly be classified as free radicals, they do appear to be associated with the general molecular system. It would be fair to say, in fact, that a large number of the interesting biological and biochemical processes under general investigation at the moment owe a considerable amount of their interest to the unpaired electrons associated with them, and hence the technique of electron resonance may be expected to have a very wide application throughout this whole field.

Before the general application of electron resonance to these different kinds of biological and biochemical systems is considered in detail, a brief review of the basic principles underlying the method will first be given. This will then be followed by a summary of the experimental techniques that are particularly appropriate in the study of such systems. It will be assumed in this treatment that the reader is not too familiar with any of the detailed work that has already been carried out by electron resonance methods in the fields of physics and chemistry, and no particular acquaintance with electronics or microwave techniques will be necessary. As a corollary, the consideration of the technical details of the spectrometer systems will only be marginal, but references will be given throughout to more advanced texts, in which full details of these subjects may be found.

§1.2

THE BASIC PRINCIPLES OF ELECTRON RESONANCE

As mentioned already, the basic feature of E.S.R. is its ability to detect and characterize the presence of unpaired electrons in a substance. The essential property detected by the technique is, in fact, the magnetic moment associated with the electron spin. Each unpaired electron has a spin angular momentum associated with it, which is actually equal to $(h/2\pi)\sqrt{[\frac{1}{2}(\frac{1}{2} + 1)]}$, this being given by the quantum number $s = \frac{1}{2}$. It should possibly be explained at

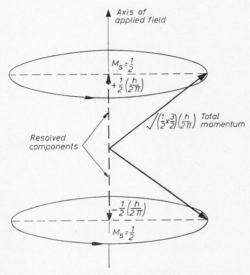

FIG. 1.1. Resolved components of total spin momentum and magnetic moment along an axis of quantization

this point that the angular momentum associated with any quantum number, such as n, is always given by $(h/2\pi)\sqrt{[n(n + 1)]}$ rather than by a simple value $(h/2\pi) \times n$. The reason for this is that the total angular momentum is precessing about a fixed axis in such a way that the permanent time-averaged value along this axis is equal to $(h/2\pi) \times n$, whereas the component perpendicular to this axis is averaged out by the precessional motion. This fact is illustrated in Fig. 1.1 for the particular case of the single unpaired electron, where the appropriate quantum number is $s = \frac{1}{2}$, as already mentioned. When the particular states of the unpaired electron are being considered, they are therefore characterized by the magnitude of the spin momentum resolved along this axis of reference and hence, for the case of the single unpaired electron, would be characterized by $M_s = +\frac{1}{2}$ or $M_s = -\frac{1}{2}$, according to the orientation of the total angular momentum, as indicated in Fig. 1.1.

The characterization of the different unpaired electrons in this way necessitates the presence of a common axis of reference, however, and in the absence of any external applied magnetic, or electric, fields, such a common axis of reference will not exist. Thus the unpaired electrons associated with the free radicals, or enzymes, present in a system standing on the laboratory bench would have a random orientation of their axes and would all have the same energy. If an external magnetic field is applied across the specimen, however, the electrons will now have a common axis of reference and will all precess so that they have a resolved component of either $M_s = +\frac{1}{2}$ or $M_s = -\frac{1}{2}$ in the direction of this applied field. General quantum conditions allow only quantum states in which the quantum numbers differ by unity; hence, the two cases of $M_s = +\frac{1}{2}$ and $M_s = -\frac{1}{2}$ are the only ones allowed for these single unpaired electrons.

It is also evident that the application of such an external magnetic field divides the unpaired electrons into two groups, those with their spins aligned parallel and those with their spins antiparallel to the direction of the field itself. It does, moreover, also give different energies to the electrons in these two groups. Those which have their magnetic moments lined up parallel to the magnetic field will have their energies reduced, whereas those lined up antiparallel to the field will have their energies increased. The reason for this can be seen by the analogy of a simple bar magnet and its behaviour in the magnetic field produced between the pole faces of an electromagnet. Normally, the bar magnet will be placed in this magnetic field so that its north pole faces the south pole of the electromagnet, and vice versa. This is, in fact, the position of stable equilibrium and a small displacement of the bar magnet from this position will be rapidly rectified, with the magnet swinging back to its original orientation. If the field of the electromagnet is completely symmetrical, however, it is possible in principle to place the small bar magnet in the field in exactly the opposite direction, i.e. with its north pole facing the north pole of the electromagnet and its south pole facing the south pole of the electromagnet. This position is, however, one of unstable equilibrium and a small displacement of the bar magnet will then cause it to swing rapidly round to the opposite orientation which is, of course, one of higher stability, and therefore also one of lower energy.

The actual magnitude of the energy of a bar magnet in a magnetic field can be written simply as $-\mu H$, where μ is the magnitude of the magnetic moment in the direction of the applied field. If we now apply this reasoning to the case of the two groups of unpaired electrons, we find that the ones with their magnetic moments lined up in the direction of the field are reduced in energy by an amount $\frac{1}{2}g\beta H$, while those lined up antiparallel to the field have their energy increased by the same amount. In this expression the '$\frac{1}{2}$' comes from the spin quantum number M_s equal to $\pm\frac{1}{2}$, the 'β' is called the Bohr magneton, and effectively converts the units of angular momentum, defined by the M_s quantum number, into units of magnetic moment, while the 'g' is an additional constant, which is necessary to allow for the fact that the magnetic moment associated with electron spin is anomalously high, and for a free spin is about twice the value expected on classical theory. The magnetic moment associated with the orbital angular momentum is, in fact, exactly that expected from a simple classical treatment, and hence an electron with one quantum of orbital angular momentum would have exactly one Bohr magneton of magnetic moment associated with it. For a completely free electron, with no orbital motion and with all its angular momentum arising from its spin, the g-value would, in fact, be 2·0023, the small correcting factor at the end coming from the interaction of the electron with its own radiation field. As a result of both of these effects, i.e. the fact that the angular momentum of the spin is only $\frac{1}{2}$, and also the fact that its magnetic moment is approximately twice what it should be, it follows that the magnetic moment associated with one unpaired electron spin is also approximately one Bohr magneton (it is in fact 1·00114 of a Bohr magneton).

It has been seen that the application of the external field across the substance containing the unpaired electrons has thus produced two groups of these electrons, and also produced an energy level splitting between them, as illustrated on the right-hand side of Fig. 1.2. The basic principle of the electron resonance technique is now to apply electromagnetic radiation at the same time, and arrange for the frequency, v, of this radiation to be such that hv, the quantum of energy associated with it, is equal to the energy difference between the two groups of electrons, i.e.

$$hv = g\beta H \qquad (1.1)$$

The experimental technique therefore involves the application of a strong homogeneous external magnetic field across the sample, and a simultaneous application of electromagnetic radiation of the correct resonance frequency. This will then supply energy to the unpaired electrons in the lower energy level, exciting them to the upper level and reversing their spins in the process. This absorption of energy by the electrons as they jump to the higher level can be detected as an actual reduction in the power of the electromagnetic radiation passing through the system, and this absorption can then be displayed on an oscilloscope screen, or in any other way that is normal in absorption spectroscopy. If quantitative figures for the

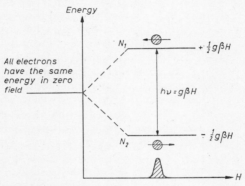

Fig. 1.2. Basic electron resonance condition
Splitting of the energy levels of a single unpaired electron under the influence of an applied magnetic field H. Resonance condition is $h\nu = g\beta H$, $\nu = 2 \cdot 8 \times 10^6 \ H$ c/s for a free electron

constants involved are substituted into equation (1.1) and the electron is assumed to have only spin momentum, it will be seen that the frequency required for the electromagnetic radiation is given by

$$\nu = 2 \cdot 8 \times 10^6 \times H \ \text{c/s} \qquad (1.2)$$

where H is the value of the magnetic field in gauss. Thus for the value of magnetic field that can be produced by ordinary electromagnets, which reach a maximum at about 10 000 gauss, the value of the resonant frequency will be about 28 000 Mc/s, or, in other words, a frequency in the radar, or microwave, region.

It may be argued, however, that since equation (1.1) is a resonance condition, it should be possible to fulfil the requirements by reducing both the frequency and magnetic field at the same time, for instance, by reducing the values quoted above by a factor of 1000 on each side, and thus carry out electron resonance experiments in magnetic fields of 10 gauss with applied radio frequencies of 28 Mc/s. Electron resonance has, in fact, been performed under these conditions, but there is, nevertheless, one very important reason why experiments are normally conducted at values of magnetic field strength and frequency as high as possible.

In fact, only half of the mechanism concerned with the interaction of the electromagnetic radiation with the electrons has been considered as yet. At the same time as the incoming radiation induces electrons in the lower level to absorb energy and jump to the higher level, it also induces electrons which are already in the higher level to fall down to the lower level, and thus emit an additional quantum of electromagnetic radiation in the process. This last process is termed 'stimulated emission', and can be considered as exactly opposite to the normal absorption process. Moreover, early in the history of radiation theory Einstein showed that the coefficients of absorption and stimulated emission were equal, and thus if there were equal numbers of unpaired electrons in the two energy levels concerned, there would be just as many electrons being driven down and emitting radiation as there were being driven up and absorbing radiation, and there would thus be no net absorption. The only reason why any microwave absorption is in fact obtained is that, under normal conditions, the unpaired electrons are always somewhat more numerous in the lower energy level than in the higher, and therefore the amount of absorption is larger than the amount of stimulated emission which is also taking place. It follows, moreover, that the difference in population between these two levels is the crucial parameter governing the strength of the absorption signal that is actually observed; hence, it is one of the most important parameters when considering the sensitivity of the spectrometer itself.

Most normal systems obey a Maxwell–Boltzmann form of statistics in which the population between the two energy levels is given by the equation

$$\frac{N_1}{N_2} = e^{-h\nu/kT} \qquad (1.3)$$

It therefore follows that if the difference between N_1 and N_2 is to be made as large as possible, the value of $h\nu$ must also be made as large as possible, and this will also apply to the value of applied magnetic field H, according to equation (1.1). It is this basic fact of energy-level population, as governed by statistical mechanics, that therefore explains why electron resonance spectrometers should always be operated at values of magnetic field and frequency that are as high as possible. In practice, most spectrometers

B

operate either at a frequency of 9000 Mc/s, which corresponds to a wavelength of 3·2 cm and is, in fact, used for marine radar (X-band), or at a frequency of 36 000 Mc/s, which corresponds to a wavelength of 8 mm and is the region of airport control radar (Q-band). The great advantage of using spectrometers in one or other of these particular wavelength regions is that all the microwave techniques and components have been very well developed for the radar applications and can thus be taken over from these fields. The magnetic field strengths corresponding to the *g*-value of 2·0 for a free electron for these two frequencies are in fact, 3300 gauss and about 13 000 gauss respectively. Before the advent of superconducting magnets, the latter field represented more or less the maximum that could be readily obtained with reasonable homogeneity from a normal electromagnet, and hence Q-band, or 8-mm-wavelength, spectrometers have been considered as the highest frequency instruments readily available. Of recent years, however, considerable interest has shifted to the higher frequency and shorter wavelength spectrometers, especially in connection with zero-field splittings, and these are also considered in some detail later.

§1.3

MAIN FEATURES OF E.S.R. SPECTRA

1.3.1   *Integrated intensity*

It was seen in the previous section that the presence of unpaired electrons produces absorption of the incoming microwave radiation and that the magnitude of this absorption is proportional to the difference in energy-level population between the two levels concerned. At a given value of frequency and temperature, this difference in population remains as a constant fraction of the total number of unpaired electrons; hence, the actual magnitude of the microwave absorption is proportional to the total number of the unpaired electrons present in the sample. It follows that the actual integrated intensity of the absorption line can be used as a measure of the concentration of the unpaired electrons present in the specimen (provided the specimen is not being "saturated" as discussed in a later section), and this kind of measurement can be of considerable interest in, for instance, the plotting of free-radical concentrations or enzyme activity during a transient phase.

There are two other parameters associated with a single absorption line, viz. the line width and the *g*-value. Both can give information of considerable value in biological and biochemical systems;

they are, therefore, briefly considered before the different factors which produce splittings of the spectra are discussed.

1.3.2 *Line width*

A broad shallow line may have the same integrated intensity as a sharp narrow line, but its width at half-height may be very different. Line width is, therefore, an additional parameter defining the absorption condition. It will be obvious that the line width is determined directly by the energy spread of the levels occupied by the unpaired electrons in question, and hence a measurement of the line width will give information on the actual interactions which the unpaired electron is experiencing and which cause its energy to be spread out.

Generally speaking, there are two main types of interaction which can produce a noticeable broadening of the electron resonance line. The first type is given the name 'spin–lattice interaction', and represents the interaction, or interactions, which can take place between the unpaired electron spin and its general surroundings, i.e. either the crystalline lattice or the rest of the molecule and molecular system, in which it is immersed. This mechanism, by which energy taken up by the spins is returned to the lattice or molecular system as a whole, must indeed be present if the actual phenomenon of resonance absorption is to continue. If it were not, electrons originally in the lower energy level would be raised to the upper level by the incoming microwave energy, until the two levels were equally populated, and then absorption would cease. If, however, this interaction is very strong, the electrons may remain in the upper level for only a very short period of time, Δt, and, if this is so, there will be a corresponding spread of energy in this upper level, as given by the uncertainty principle relation:

$$\Delta E \cdot \Delta t = \frac{h}{2\pi} \tag{1.4}$$

Hence considerable broadening of the resonance line might be produced. For this reason, one of the most realistic ways of measuring such an interaction with the rest of the molecular system is in terms of the spin–lattice relaxation time, which can be defined as the time taken by the electron spin system to lose $1/e$ of the energy it has received from the microwaves. It then follows that a strong spin–lattice interaction will produce a short spin–lattice relaxation time, τ, and a broadening of the absorption lines. The actual broadening produced, as measured in frequency units, can be related directly to the inverse of the relaxation time defined in this way by:

$$\Delta v = \frac{1}{2\pi . \tau} \qquad (1.5)$$

The other types of broadening interaction which occur can be grouped together under the general term of 'spin–spin interaction', and contain all the mechanisms whereby the spins can exchange energy amongst themselves, rather than giving it back to the lattice, or molecular, system as a whole. Such interactions will not help in the general establishment of thermal equilibrium, as with spin–lattice relaxation, but can produce a broadening of the resonance line, both from the direct action of the spins on each other and from the smaller lifetime of spin states, which they can also induce. One of the major interactions of this type is the ordinary dipole–dipole interaction, which can be considered equivalent to the classical interaction of two bar magnets. Thus, each of the unpaired electrons in the specimen will react not only to the externally applied magnetic field, but also to the magnetic field produced by the other unpaired electrons. If the concentration of the electrons in the specimen is not too low, so that the neighbouring unpaired electrons are not very far apart, this additional field can have quite a significant magnitude. In fact, for single crystals of concentrated magnetic salts, additional fields of the order of hundreds of gauss can be produced from the neighbouring magnetic moments. However, in studies of dilute solutions, such as are normally found in biological and biochemical applications, the concentration of the unpaired electrons is very much smaller and their effect on one another is thus very much reduced. Detailed consideration of both the spin–lattice and spin–spin interactions is also given in the introductory volume to this series of monographs,[1] and in other general books on electron resonance, such as *Spectroscopy at Radio and Microwave Frequencies*,[2] and further discussion of the different mechanisms associated with these two types of interactions may be found there.

One point that should be noted, however, is that these different types of interaction not only produce different magnitudes of broadening in the absorption lines, but also often affect the shape of the absorption line; thus the normal spin–spin dipolar interaction will produce a line shape of a Gaussian form, as illustrated in Fig. 1.3(a). On the other hand, the line shape produced by most of the spin–lattice interactions, and also by the exchange or averaging effects as they occur in solutions, takes on a Lorentzian form, as illustrated in Fig. 1.3(b). It is seen that this particular line shape is narrower in the centre, but spreads out with more intensity in the wings, than that of the Gaussian line. It may also be seen from the figure that the actual widths of the line, at half-power level, are different for the two cases, and in order to express this in a quantitative fashion, the lines are normally expressed by an

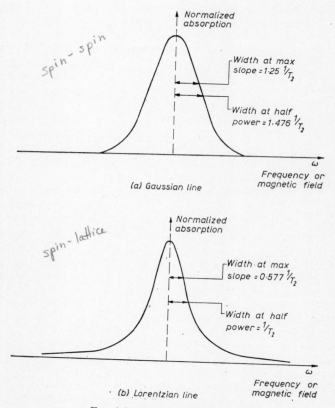

FIG. 1.3. Absorption line shapes

equation defining their shape and written in a normalized form such that

$$\int_0^\infty g(\omega - \omega_0)\, d\omega = 1 \tag{1.6}$$

The actual equations defining the Gaussian and Lorentzian shapes are then given by

$$g(\omega - \omega_0) = \frac{T_2}{\pi} \cdot e^{-(\omega - \omega_0)^2 T_2{}^2/\pi} \qquad (1.7)$$

$$g(\omega - \omega_0) = \frac{T_2/\pi}{1 + (\omega - \omega_0)^2 \cdot T_2{}^2} \qquad (1.8)$$

A useful parameter can then be defined from this line-shape function. It is denoted by T_2 and is defined by the equation

$$T_2 = \pi \cdot g(\omega - \omega_0)_{max} \qquad (1.9)$$

The narrower the line, the larger the value of the maximum of the line-shape function, and hence the larger the value of T_2. For a Lorentzian line, the width at half power is actually equal to $1/T_2$, as shown in Fig. 1.3(*b*), whereas the width at half power for a Gaussian line is $1.476 \times 1/T_2$. The value of T_2 can also be taken as a measure of the interaction producing the line width, and it is often termed the 'transverse' or 'spin–spin' relaxation time, since it is directly related to the time taken for the re-establishment of equilibrium within the spin system.

The way in which these line-shape measurements and studies can be used to give information on biological and biochemical specimens will be discussed in detail in Chapters 5, 6 and 7, but these introductory descriptions may serve to show that even a rapid glance at the shape of the absorption line can give definite information on the interactions taking place within the unpaired electron system.

1.3.3   *g-values*

The other parameter which may be associated with a single absorption line is its actual resonance position. Reference back to equation (1.1) shows that, if the microwave frequency is held constant, the only two parameters that can vary are the value of the externally applied magnetic field and value of the *g*-factor. Hence, a measurement of the value of the field at which resonance occurs determines the value of the *g*-factor associated with the particular unpaired electron or the atom or molecule in which it is residing.

As already seen, an electron with no orbital angular momentum has a *g*-value equal to that of the free electron spin, i.e. equal to 2·0023. A large number of free radicals, in which the electron moves in a widely delocalized molecular orbital, do, in fact, have *g*-values very close to this magnitude, indicating very little coupling to any orbital motion associated with an atomic orbital. A measurement of a *g*-value very close to the free-spin value therefore indicates either a highly delocalized molecular orbital or a state in which the electron has no orbital motion associated with it. This can be useful information, but the magnitude of the *g*-value itself then gives no further information on the state of the electron or its associated atom.

If, on the other hand, the electron is moving in an atomic orbital associated with a single atom, it may possess considerable orbital angular momentum and this will shift the *g*-value away from that appropriate to the free spin. The basic reason for this shift in *g*-value is that the relationship between magnetic moment and angular momentum is different for spin motion and for orbital motion. In the case of orbital angular momentum the relationship is that given directly by the simple classical analogy as already explained in §1.2, whereas the magnetic moment associated with spin angular momentum is more or less twice what it should be. In the case of an electron associated with a free atom, which has no external magnetic or electric fields acting on it, the resultant *g*-value can be derived directly in terms of the quantum numbers defining the total spin and orbital magnetic moments, S, L and J. The *g*-value is, in fact, then identical with what is known as the Landé splitting factor and is given by the equation:

$$g = 1 + \frac{J(J+1) + S(S+1) - L(L+1)}{2J(J+1)} \qquad (1.10)$$

If the unpaired electron is not associated with such a free atom, however, but with an atom contained within a solid crystalline lattice or other molecular structure, as is nearly always the case in any biological or biochemical specimen, then quite strong internal electric fields, which arise from the structure of the molecule itself, will also be acting on it. These electric fields act on the orbital states of the atom in question and can very radically alter their energies, and, as a result, the simple Landé factor can no longer be applied, and the calculation of the *g*-value, appropriate to the particular case, becomes more complicated. The details of the way in which the *g*-values can be calculated for different transition group atoms in different situations are to be discussed specifically in a later volume of this series of monographs.[3] The particular cases which are of interest in biological compounds, such as copper, molybdenum and iron, are also discussed later in Chapter 5, and the only thing which may be noted in this introductory section is that the shift of the *g*-value away from 2·0, and its angular variation, can give very detailed information on the type of chemical bonding associated with the atom or radical in question. The measurement of these *g*-values can give very detailed information on all aspects of the chemical bonding within the molecular structure, including the existence of covalent bonding within particular complexes.

The above three parameters have all been defined and discussed for the case of a single absorption line, and their actual measurement and determination can often give valuable information in the ways indicated. On the other hand, it would probably be fair to say that, if electron resonance produced only a single absorption line of this type when systems containing unpaired electrons were being studied, then a large number of its applications would never have been discovered and it would not have proved the versatile tool that it now is. Most of these other applications come from the splittings that can also be produced in the spectrum and which give further information on both the energy states and the location of the unpaired electrons. There are two main kinds of splittings that can be produced in an electron resonance spectrum, both of which have major applications in the study of biological or biochemical specimens. The first of these, known as 'the electronic splitting', arises when the atom or molecule concerned possesses more than one unpaired electron. The second, termed 'hyperfine splitting', arises when the unpaired electron interacts with the magnetic moment of the nuclei embraced within its orbital. The two kinds of splitting will now be considered in turn, and the main effects which they have on the spectrum will be outlined.

§1.4

ELECTRONIC SPLITTING

1.4.1   *Atoms or molecules with more than one unpaired electron*

In the discussion on the basic principles of electron resonance given in §1.2, it was assumed that the compound under study possessed only one unpaired electron per atom, or molecule, which is indeed generally the case for free-radical specimens. However, it is possible for such molecular species to possess two unpaired electrons per molecule, and triplet states in which these two unpaired electrons have been formed by incoming radiation play, in fact, a very significant part in biological studies. It is also the case that many of the transition group atoms have several unpaired electrons associated with the inner orbitals, which are screened by the outer valency electrons. The ease with which such atoms can change their valency is associated with the different number of electrons that can be placed in these inner orbits; it is, of course, also the reason why they often play such a significant part in such processes as enzyme activity. The situation which arises when there is more than one unpaired electron per atom or molecule can probably be best approached by considering the simpler case of the triplet state system, in which there are two unpaired electrons associated with a molecular orbital. Once the features of this case have been outlined, the somewhat more complex situation which arises in the transition group atoms can be discussed.

1.4.2 *Electronic splitting in organic molecules*

The two unpaired electrons associated with the triplet state in an organic molecule can be considered as possessing only spin angular momentum, and their two momenta will, therefore, add vectorally to give a total spin quantum number of $S = 1$. This

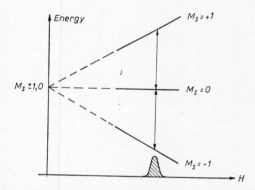

Fig. 1.4. Splitting of the energy levels associated with two unpaired electrons with $S = 1$ in the absence of internal fields

combined angular momentum and magnetic moment will now orientate as a unit in any applied magnetic field, and it follows that there are now three different orientations which are possible, corresponding to the resolved quantum numbers of $M_s = +1$, 0 or -1.

If the two unpaired electrons are not also subjected to an electric field, then all these three orientations of the total spin will have equal energies in zero applied magnetic field, and their energies will diverge, as shown in Fig. 1.4, when the external field is applied. It is seen that the level corresponding to the zero component along the field axis will, in fact, be unchanged in energy, as would be expected, whereas the other two have their energies changed at twice the rate corresponding to the levels represented in Fig. 1.2, where the total magnetic moment corresponds to only one electron spin. If electromagnetic radiation of frequency corresponding to the energy difference between the $+1$ and the 0 level or the 0 and the -1 level is now applied, absorption will occur in exactly the same way as previously discussed. In fact, it is clear from Fig. 1.4 that the resonance condition between the two sets of levels occurs at exactly the same field for each of the two allowed transitions, so that, although two transitions are actually taking place, only one absorption line is observed. The selection rule in all these cases is that the quantum number can only change by 1 and thus, to a first order, the change $\Delta M_s = \pm 2$, which would produce a transition across the outside levels at half the field strength, is

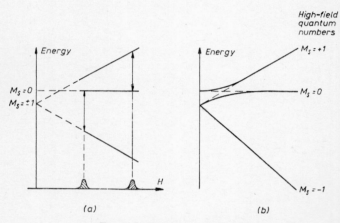

Fig. 1.5. Electronic splitting

(a) Schematic diagram showing how a zero-field splitting produces two absorption lines

(b) Actual divergence of energy levels in region where axis of quantization is indefinite when magnetic field is perpendicular to crystalline axis

not allowed. There is, therefore, no splitting of the absorption line under these conditions.

In practice, however, molecules containing the two unpaired electrons will not be in a region of zero electric field since the crystal structure in which it is sitting, or the general molecular configuration of a non-crystallized medium, will have produced quite strong internal electric fields acting upon the unpaired electron. These internal electric fields have an effect, known as the Stark effect, on the energy levels of the electrons, separating them into two groups, as illustrated on the left of Fig. 1.5(a). It is, in fact, a general theorem of the Stark effect that an axially symmetric electric field cannot separate energy levels which have the same quantum number but are of opposite sign, and can only separate those corresponding to different quantum numbers. Thus the $M_s = \pm 1$ levels still remain degenerate in the presence of the molecular electric field, but are separated from the $M_s = 0$ level. If electromagnetic radiation of frequency v is now applied to the specimen, it is evident that the absorption no longer takes place at the same value of the applied magnetic field for the two transitions already considered. The two transitions now require different values of field to produce the necessary energy difference as indicated on the right-hand side of Fig. 1.5(a); as a result, two absorption lines are obtained well separated from each other. The splitting thus produced is termed an electronic splitting since it arises from an interaction between the unpaired electrons present in the atom, or molecule. The magnetic field separation can be taken as a direct measure of the actual energy separation between the different component levels produced by the internal electric field. Measurements on such electronic splittings can, therefore, give very useful information on the internal interactions taking place within the molecule itself.

It should be mentioned that this description of the energy-level separations has been somewhat simplified to clarify the lines of argument. A slightly more refined treatment, in which the second-order effects are taken into account, especially in the lower-field region, where they are more important, would give an energy-level picture as shown in Fig. 1.5(b) for the case when the applied magnetic field is perpendicular to the axis of the internal electric field. In this case, the energy levels never actually cross but one merges into the other, although the final result on the high-field side is exactly as predicted by the simpler treatment applied in Fig. 1.5(a). These points will be taken up again in more detail when applications of zero-field splittings to biochemical compounds are considered in a later section.

This somewhat curious behaviour of the energy levels arises from the competition between the internal electric field and the external magnetic field, the former acting as the axis of quantization at low fields, and the latter at high fields. This competition also produces a noticeable angular variation in the spectrum when the direction of the applied magnetic field is varied with respect to the molecular, or crystalline, axis. Thus the value of the electronic splitting which is measured will depend on the orientation which the applied magnetic field makes to the crystalline, or molecular, axis, as indicated in Fig. 1.6. If a single crystal is being studied, this variation in the electronic splitting can be easily followed and can be systematically plotted against the angle between the magnetic field and the crystalline axis; considerable information on the symmetry and magnitude of the internal molecular field can thus be obtained. If, however, the specimen to be studied exists in a liquid, or amorphous state, then only an average of all the possible electronic splittings will be observed, and this

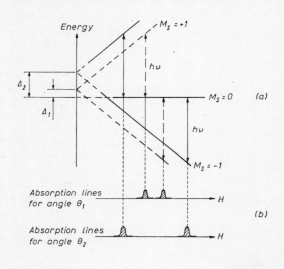

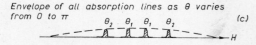

FIG. 1.6. Angular variation of electronic splitting associated with triplet state $S = 1$

(a) Energy level splittings for two different angles.
  Δ_1 = Effective zero-field splitting for angle θ_1
  Δ_2 = Effective zero-field splitting for angle θ_2.
(b) Resultant absorption lines
(c) Envelope of all possible absorption lines showing weak smeared out signal

can produce a broad smeared-out absorption, as indicated in Fig. 1.6(c), which may be quite undetectable. This smearing out of the absorption by a large angular variation in the spectrum is a feature which can quite often occur in electron resonance of liquid or amorphous substances, and care must always be taken when making deductions from negative results. The initial observation of electron resonance from excited triplet states was only obtained, in fact, when single-crystal studies were carried out, although more recent work has shown that they can be detected in solution as well.

1.4.3 *Electronic splitting in transition group atoms*

In the above considerations of the electronic splitting, the case of a triplet state has been taken where the two unpaired electrons are associated with a delocalized molecular orbital. It is possible, however, for single atoms to have more than one unpaired electron associated with them, and indeed this is quite common for the transition group atoms in particular. In the case of the first transition group, for instance, the 3d shell of electrons fills up after electrons have entered the 4s shell, and hence it is possible to obtain atoms, or ions, with numbers of electrons in the 3d shell varying from one to ten, and it is thus possible for up to five unpaired electrons to exist in a given atomic shell. The case of five unpaired electrons will actually occur for the ferric and manganous ions which both have five electrons in the 3d shell, and these can all line up in the same direction to give a total electron spin quantum number of $S = \frac{5}{2}$. These particular ions are, in fact, of some considerable interest in biological and biochemical studies, since the ferric ion plays an important role in such molecules as haemoglobin. On the other hand, manganous ions are often found as impurities in aqueous solutions, and hence the spectrum they produce often has to be eliminated from others which are being investigated. It may, therefore, be helpful if the particular case of the electronic structure associated with these two ions is considered in some detail, as a second example of what is meant by electronic splitting in a typical electron resonance spectrum.

It should be pointed out that the total electron spin quantum number $S = \frac{5}{2}$ can be obtained only for these two ions, which are both in a ^6S spectroscopic state, if the ions are ionically bound to their nearest neighbours and strong covalent bonding does not exist. It will be seen later than such strong covalent bonding can alter the number of unpaired electrons associated with the ferric or manganous ion, but for the present example ionic-type bonding will be assumed with the corollary that each of the five 3d electrons enters a separate orbital and that all thus couple with aligned spins to give $S = \frac{5}{2}$. This ^6S spectroscopic state is, in fact, an orbital singlet state that has no orbital angular momentum associated with it. Thus, to a first approximation, the magnetic moments can be considered as due entirely to the spin angular momentum, and hence the g-value will also be very close to 2·0. If the ferric or manganous ions existed in a free state and were not subjected to the electrostatic fields inside the crystal or molecule, then all the possible orientations of the $S = \frac{5}{2}$ total spin vector would have the same energy, and hence all would be degenerate on an energy-level diagram, as shown in Fig. 1.7(a). The application of an external magnetic field would then cause the spins to take up different

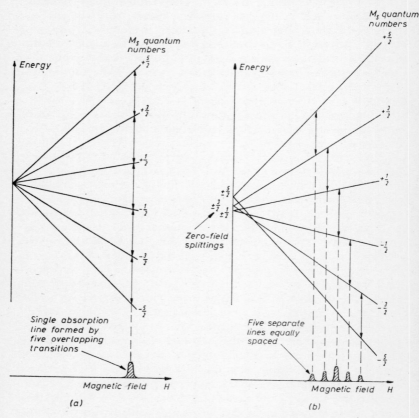

FIG. 1.7. Electronic splitting for Mn^{++} ion

(a) Divergence of different component levels with applied magnetic field. If there is no internal field, all of the six M_s levels have the same energy in zero field
(b) Divergence of different component levels from zero field splittings produced by the internal fields

quantized orientations with respect to this field, the possible orientations varying from those with a component of $M_s = +\frac{5}{2}$ in the direction of the field through those with orientations corresponding to $M_s = +\frac{3}{2}; +\frac{1}{2}; -\frac{1}{2}; -\frac{3}{2}$ to the other extreme of $-\frac{5}{2}$

Moreover the energies of these different orientations will diverge with the application of the applied magnetic field, by amounts proportional to the resolved spin quantum number, and hence the divergence of the energy levels with the application of the applied field will be as shown in Fig. 1.7(a). Transitions from one state to another must again obey the normal quantum rule $\Delta M_s = \pm 1$ so that only those indicated by the arrows are allowed, and it is evident that for a given incoming microwave frequency all the possible transitions will take place at the same value of the externally applied field, and hence a single absorption line will be produced from the five overlapping transitions, in a very similar way to that discussed for the three levels of Fig. 1.5(a).

If, however, the ferric, or manganous, ion is present in a crystal or molecule, it will be acted on by the internal electrostatic fields, and hence, a Stark effect will occur and the energy levels will be separated, as in the case of a triplet state. The magnitude of these separations will again depend on the symmetry and strength of the internal fields, but the same general condition will apply, i.e. that an axially symmetric electric field, by itself, cannot remove the twofold degeneracy associated with states of the same resolved quantum number. The action of the internal field will therefore be as indicated on the left-hand side of Fig. 1.7(b), and zero-field electronic splittings will be produced between the $\pm\frac{1}{2}$, the $\pm\frac{3}{2}$ and the $\pm\frac{5}{2}$ levels, as indicated. Application of the external magnetic field will now remove the twofold degeneracy of these spin levels, and the levels will diverge in the way indicated in Fig. 1.7(b). At high enough values of the external magnetic field they will be arranged in an energy order from $+\frac{5}{2}$ to $-\frac{5}{2}$, as indicated. The allowed transitions between them no longer occur at the same value of the external magnetic field, however, and as the example given in Fig. 1.7(b) indicates, five separate absorptions will now be obtained. Moreover, the separation between these is a direct reflection of the separation produced in zero magnetic field by the electrostatic internal fields, and can therefore be used as a measure of this quantity, as in the case of a triplet state.

The above description of the variation of the energy levels with strength of applied field has again ignored the second-order effects and competition between the electric and magnetic fields for the effective axis of quantization. The behaviour of the energy levels as they go back into low fields is more complicated than that actually shown, but again, these effects are only important in the low-field-overlap region. As in the case of the triplet state, there is also a strong angular dependence of the electronic splitting and the position of the five transitions will thus vary markedly with the angle which the external magnetic field makes to the axis of the internal electrostatic fields. This variation will have the form of a normal dipolar interaction to a first approximation, and this is characterized by a $(3 \cos^2 \theta - 1)$ variation. This can be plotted out

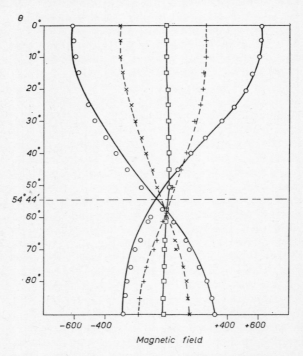

FIG. 1.8. Angular variation of electronic splitting
Experimentally observed splittings for Mn^{++} in manganese fluosilicate
at 3 cm wavelength

experimentally, and the $(3 \cos^2 \theta - 1)$ variation is demonstrated quite effectively as shown in Fig. 1.8. In this figure the positions of the five lines, corresponding to these five electronic transitions, are plotted for a manganous salt on which measurements were taken at 3·2 cm wavelengths, and magnetic fields of around 3000 gauss. These magnetic fields are, in fact, not quite large enough to reduce all the second-order effects mentioned above to a negligible factor, and hence there is some departure in the experimentally measured points from a simple $(3 \cos^2 \theta - 1)$ variation. The main features of this can be clearly seen in the figure, however, and in particular, the collapse of the splitting to nearly zero at a value of

$\theta = 54° 44'$, which is the angle which makes $(3 \cos^2 \theta - 1)$ equal to zero.

Full details of the way in which this electronic splitting can be considered theoretically, and the way in which the second-order and other effects can be allowed for, may be found in the introductory volume to this series of monographs.[4] The information that can be obtained from such zero-field electronic splittings, as they are measured in metallic-organic compounds, is also considered in much more detail in Chapter 6.

§1.5

HYPERFINE SPLITTINGS AND THEIR ORIGINS

1.5.1 *Interaction with nuclear magnetic moments*

The other kind of splitting that can occur in electron resonance spectra, and which can give a large amount of information on the energy and location of the unpaired electron, and on the properties of the molecule in general, is the hyperfine splitting. This arises from the interaction of the unpaired electron with the magnetic moment of the nucleus of any atom with which its orbit may be associated. If the unpaired electron is confined to an atomic orbital associated with only one atom, then the interaction will be with the magnetic moment of the nucleus of this particular atom. Although all nuclei do not have spins and magnetic moments, those that do include a large number of atoms which are normal constituents of biochemical compounds, such as hydrogen and nitrogen. In the case of unpaired electrons moving in molecular orbitals which embrace several atoms, the hyperfine splitting may then be produced by interaction with several different nuclei and quite complicated patterns may be obtained. The simpler case of an interaction with one nucleus will, therefore, be considered first, and then this will be extended to the more general case normally occurring in free-radical spectra.

1.5.2 *Hyperfine interaction with only one nucleus*

The case of the manganous ion can again be taken as an illustration of this point, since the Mn^{55} nucleus does possess a magnetic moment and nuclear spin, the value of the spin being $I = \frac{5}{2}$. It, therefore, follows that the unpaired electron will not only experience the externally applied magnetic field, but also interact with the field produced by the magnetic moment of the nucleus itself. Although the magnetic moments of nuclei are very small, being about 2000 times smaller than that of a single electron, the magnetic fields which they produce at the site of their own electrons can

c

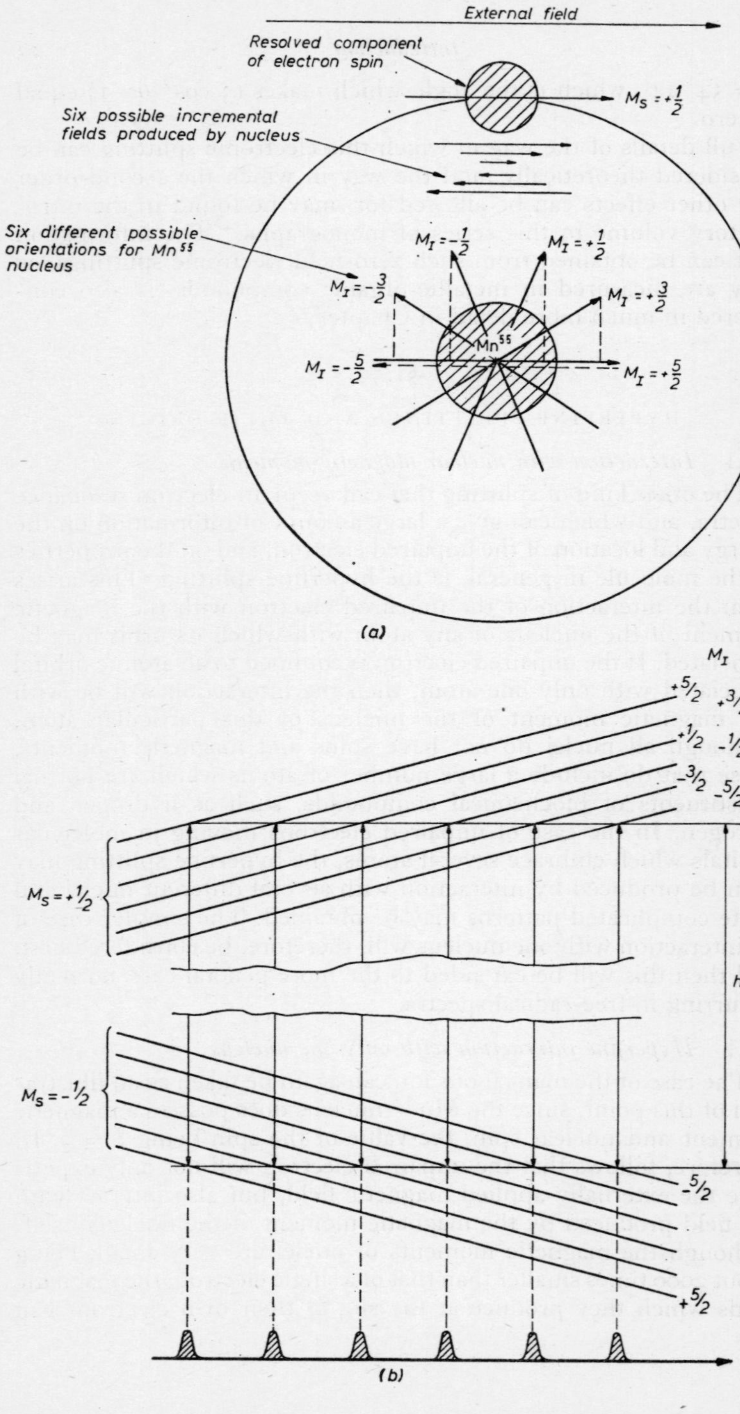
External field
Resolved component of electron spin
$M_S = +\frac{1}{2}$
Six possible incremental fields produced by nucleus
Six different possible orientations for Mn^{55} nucleus
$M_I = -\frac{1}{2}$
$M_I = +\frac{1}{2}$
$M_I = -\frac{3}{2}$
$M_I = +\frac{3}{2}$
Mn^{55}
$M_I = -\frac{5}{2}$
$M_I = +\frac{5}{2}$
(a)
M_I
+5/2
+3/2
+1/2
-1/2
-3/2
-5/2
$M_S = +1/2$
$h\nu$
$M_S = -1/2$
-5/2
+5/2
(b)

nevertheless be quite considerable, since the separation involved is also extremely small. Thus a crude calculation, using concepts of ordinary classical physics, would predict that a magnetic field of about 100 gauss will be experienced by a typical electron moving around a nucleus which has a magnetic moment of about 1 nuclear magneton. A splitting of 100 gauss magnitude will be very readily observed in the absorption spectrum, since spectrometers can now resolve splittings of 100 milligauss or less.

The way in which this interaction affects the energy levels of the unpaired electrons may be illustrated as in Fig. 1.9(*a*). The particular case in which the total electronic spin of $S = \frac{5}{2}$ has a resolved component of $M_s = +\frac{1}{2}$ along the direction of the applied magnetic field may be considered first. As well as experiencing the magnetic field which is applied externally, this electronic spin also experiences the magnetic field due to the nuclear magnetic moment. In the case of the manganous nucleus the nuclear spin is $\frac{5}{2}$, and hence it is possible for the nucleus to take up six different orientations with respect to the field direction, these corresponding to the successive resolved components of M_I equal to $+\frac{5}{2}$, $+\frac{3}{2}$, $+\frac{1}{2}$, $-\frac{1}{2}$, $-\frac{3}{2}$, and $-\frac{5}{2}$. These six different orientations are an example of the general relation that any quantum number, such as I, can be resolved along a specified axis of quantization to give $(2I+1)$ resolved components which differ by an integer from one another. The six different orientations in the case of manganese are illustrated at the centre of Fig. 1.9(*a*), and in turn, these produce six incremental magnetic fields at the site of the electron spin. These six incremental fields will all have different components resolved along the direction of the external magnetic field, as indicated, and, as a result, the total magnetic field experienced by the electron spin will have one of six different possible values, the particular value depending on the orientation of the particular nucleus around which the electron is orbiting. Since the energies associated with these six different nuclear orientations differ only by a very small amount, there will be an equal number of nuclei in all of the six orientations at any normal temperature and hence there will be an

---

Fig. 1.9. (*facing*) Hyperfine interaction between the unpaired electrons and the manganese nucleus

(*a*) Schematic diagram of interaction between the different orientations of the manganese nucleus and the $M_s = +\frac{1}{2}$ component of the total electron spin

(*b*) Resultant hyperfine splittings of the two electronic levels showing how the hyperfine lines are produced

Note. The horizontal magnetic field scale in this figure is much larger than that in Figs. 1.5–1.7

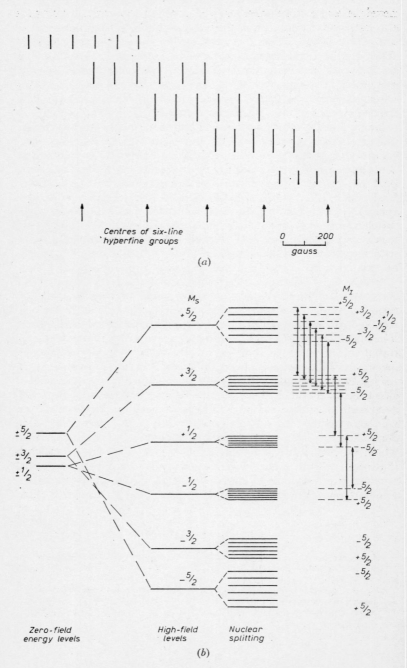

Centres of six-line hyperfine groups

0   200
gauss

(a)

M_S

$+\frac{5}{2}$

$+\frac{3}{2}$

$+\frac{1}{2}$

$-\frac{1}{2}$

$-\frac{3}{2}$

$-\frac{5}{2}$

M_I
$+\frac{5}{2}$ $+\frac{3}{2}$ $+\frac{1}{2}$
$-\frac{1}{2}$
$-\frac{3}{2}$
$-\frac{5}{2}$

$+\frac{5}{2}$
$-\frac{5}{2}$

$+\frac{5}{2}$
$-\frac{5}{2}$

$-\frac{5}{2}$
$+\frac{5}{2}$

$\pm\frac{5}{2}$
$\pm\frac{3}{2}$
$\pm\frac{1}{2}$

$-\frac{5}{2}$
$+\frac{5}{2}$
$-\frac{5}{2}$

$+\frac{5}{2}$

Zero-field   High-field   Nuclear
energy levels   levels   splitting

(b)

equal number of unpaired electrons experiencing the six different values of total magnetic field.

These six different values of field will therefore produce a splitting of the original electronic energy levels, as indicated in Fig. 1.9(*b*). If microwave radiation of constant frequency is now fed to the specimen, as in a normal electron resonance spectrometer, absorption lines will be obtained for six different field values as indicated, the normal selection rule for all these transitions is that $\Delta M_I = 0$, since the nuclear spin does not change its orientation during the electronic transition. It is thus seen that six hyperfine lines are now produced in place of the single electronic absorption, and these six lines of equal intensity reflect the $(2I+1)$ different orientations of the nuclear spin of $I = \frac{5}{2}$ of the manganese nucleus. This very direct relation between the value of the nuclear spin itself, and the number of hyperfine components that are observed in the E.S.R. spectrum, was used in the early resonance work as a very effective way of determining unknown nuclear spins. Such hyperfine patterns can now be used as very direct probes for the quantitative and qualitative analysis of the specimens concerned. The magnitude of the separation between the hyperfine components is also of interest, since this is a direct measure of the strength of the interaction between the unpaired electron and the atomic nucleus. This is governed, in turn, by the nature of the chemical bonding and the detailed distribution of the wave functions concerned, and can therefore give very useful information on both molecular bonding and structure. Details of this are considered in the appropriate sections in Chapters 4 and 5.

This hyperfine interaction between the unpaired electron and the nucleus will take place whether there is only one unpaired electron in the molecule or several, since the interaction is between the total electron spin and the nuclear magnetic moment in question. Thus the electron resonance absorption spectrum which will actually be observed from single crystals of divalent manganese will consist of five groups of six equally intense and equally spaced hyperfine lines, each of the electronic transitions discussed in the preceding section now being split into six hyperfine components. A typical absorption spectra as observed in a hydrated manganous salt is shown in Fig. 1.10(*a*), and the six hyperfine lines associated with each electronic transition can be clearly seen. The detailed

FIG. 1.10. (*facing*) E.S.R. spectrum of manganese
(*a*) Observed spectrum showing five groups of six hyperfine lines
(*b*) Energy level diagram showing the production of the five groups of six hyperfine components

energy-level diagram showing how these hyperfine components arise on the electronic transitions, is shown in Fig. 1.10(*b*).

Other examples of such hyperfine patterns from electrons

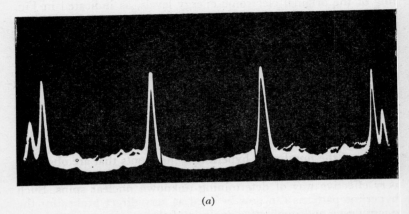

(*a*)

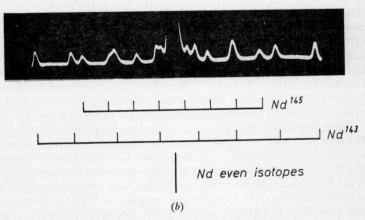

(*b*)

FIG. 1.11. Hyperfine splittings from transition group ions

(*a*) Spectrum obtained from Cu^{++} ions. Four lines are obtained from each of the Cu^{63} and Cu^{65} isotopes, as shown

(*b*) Spectrum observed from neodymium salt. The large central line is produced by the isotopes with no nuclear spin, whereas the two groups of eight lines come from the Nd^{143} and Nd^{145}, as indicated

associated with single atoms are shown in Fig. 1.11. Fig. 1.11(*a*) shows the hyperfine pattern that is obtained from a crystal containing divalent copper ions. Copper has two isotopes, Cu^{63} and Cu^{65}, both of which have a nuclear spin of $I = \frac{3}{2}$, and hence both produce a four-line hyperfine pattern. The magnetic moments of the two isotopes are slightly different, however, and thus the two patterns do not quite overlap and the separation between the two is clearly seen on the outer lines. The outermost lines come from the Cu^{63}, which has slightly less abundance than the Cu^{65}. Fig. 1.11(*b*), on the other hand, shows the hyperfine pattern observed from a salt containing neodymium. This particular spectrum is quoted as an example since it was the first in which electron resonance was used to measure a previously unknown nuclear spin; further it illustrates the kind of complexity that can be observed, even when the electron is only interacting with ‹ ne nucleus. The large central line comes from the even isotopes of the neodymium, which have no nuclear spin, or magnetic moment, and therefore produce no hyperfine pattern. It can be seen, however, that as well as this large central line, which goes off the scale, there are also two groups of eight lines, each line within the group having the same intensity. Moreover, the relative intensity of these two groups of hyperfine lines is equal to the ratio of the abundance of the Nd^{143} and Nd^{145} isotopes, and hence the nuclear spins of both of these can be immediately determined as equal to $\frac{7}{2}$. The ratio of the separation between the lines in the two groups also immediately gives the ratio of their nuclear magnetic moments. These nuclear properties themselves are not normally those which are of immediate interest to the biochemist, or biophysicist, however, and the particular parameter of interest in this connection is the actual splitting between the hyperfine components and the way in which the splitting varies with angle. This kind of information not only allows identification of the active centre in the particular compound, but can also give very direct evidence on the nature of the chemical bonding and the possible activities to be associated with different groups embraced within the orbital of the unpaired electron.

1.5.3 *Angular variation of hyperfine pattern*

All the three examples quoted above have been for the case of spectra obtained from single crystals. It was noted, in the last section, that the electronic splitting depended markedly on the angle between the applied field and the molecular axis and, unless single crystals were studied, the absorption would be smeared out beyond detection by the randomizing effect produced by the different orientations. In quite a large number of cases, the hyperfine structure is also angularly dependent, and hence can also be smeared out and masked in this way if solutions, or amorphous or polycrystalline substances, are studied. On the other hand, not all of the interaction between the unpaired electrons and the nuclei is dependent on angle, and the hyperfine interaction can, in general, be divided into two parts. The first part, known as the 'dipole–dipole' term, is that expected classically from the interaction between two small bar magnets; it has the same angular variation as the electronic splitting discussed above, that is, it varies as $(3 \cos^2 \theta - 1)$ with the angle, θ, between the direction of the applied magnetic field and the direction joining the electron and nucleus. As well as this angularly dependent term, however, there is also a second term, known as the 'Fermi' or 'contact' term, which is quite independent of angle, and is directly proportional to the magnitude of the electron wave-function that exists at the site of the nucleus. It is only possible for electrons moving in s type orbitals to have any finite intensity at the site of the nucleus itself, and hence this Fermi or contact term, which produces an isotropic hyperfine structure, will arise entirely from the s nature of the orbital of the unpaired electrons. This interaction is not averaged out in solutions or polycrystalline samples, however, since it will be of the same magnitude whatever the orientation of the molecule or crystallite in the applied magnetic field. This fact is of great importance when biological samples are studied, since they often need to be investigated in solution, and as close to *in vivo* conditions as possible. In such circumstances the angular dependent terms in the hyperfine splitting are normally averaged out and not observable, but the isotropic interactions still remain and can often be identified.

Two examples of these are shown in Fig. 1.12. The first of these is the spectrum observed from a solution containing manganous ions, which have already been discussed in some detail above. It has been seen that a single crystal containing these ions will give five sets of fixed hyperfine lines, and these five groups will move rapidly with angle. The central group, however, remains effectively stationary, and in solution the random orientation of the molecules averages out the electronic splittings, and the four outer groups of the hyperfine lines are thus removed. The central group can still be observed, however, with its well-resolved isotropic hyperfine pattern. Such a pattern is shown in Fig. 1.12(*a*), and is often to be found present in buffer solutions, used in the preparation of biochemical specimens, since manganese is a fairly common impurity and electron resonance techniques can detect it down to very low concentrations.

Fig. 1.12(*b*) gives another example of a spectrum observed from solution, in this case from an enzyme system containing copper ions. The copper present in this particular enzyme has an electron resonance spectrum with a g-value that varies with angle, and the reasons for this, and the information that can be obtained from it, are summarized in detail in Chapter 5. Here we may note that a variation in *g*-value, which is effectively also a variation in the magnetic field required for resonance, produces a general spread of the absorption line over a range of the externally applied field.

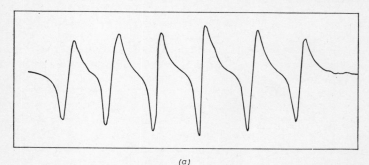

(a)

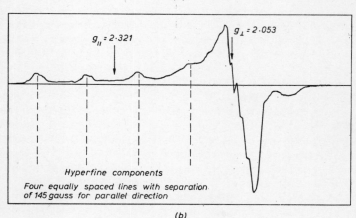

$g_{||} = 2.321$

$g_{\perp} = 2.053$

Hyperfine components
Four equally spaced lines with separation of 145 gauss for parallel direction

(b)

FIG. 1.12. Hyperfine patterns from ions in solution

(a) Mn^{++} in aqueous solution
(b) Hyperfine pattern from enzyme containing divalent copper. Four equally spaced hyperfine lines with separation of 145 gauss are observed, centred on the $g_{||}$ position

Note. These, and the spectra in Figs. 1.13 and 1.14 are first derivative tracings

One end of this spread will then correspond to the *g*-parallel, which is the *g*-value obtained when the external magnetic field is parallel to the axis of the molecular electric field, while the other end corresponds to the *g*-perpendicular direction. If axial symmetry exists, there will be approximately twice as many ions approaching the perpendicular direction as the parallel direction, and hence the two can be distinguished by virtue of the larger signal that occurs in the perpendicular direction. In some cases, the splitting between the hyperfine components also varies markedly with angle, and a wide splitting may be obtained in the parallel direction, whereas a very much smaller one is observed in the perpendicular direction. This is the case of the spectrum shown in Fig. 1.12(*b*), where the well-resolved four-line hyperfine pattern can be readily identified, even although the enzyme concerned is present in a liquid state, and all angular variations have been averaged out by the tumbling motion of the solvent molecules.

1.5.4 *Hyperfine pattern from more than one nucleus*

In the cases considered above, the unpaired electron has been assumed to be interacting with only one nucleus, e.g. either with the manganese or with the copper nucleus. In a large number of cases met in biochemical and biological investigations, however, the orbit of the unpaired electron is distributed over several atoms, and hence it is possible for the unpaired electron to interact with several nuclei at once. In this event, the hyperfine pattern reflects the interaction with these different nuclei, and hence can become quite complicated. In practice, however, such a multi-nuclear interaction normally takes one of two basic forms; either the unpaired electron interacts equally with a number of similar nuclei, as is often the case in free radicals, or the unpaired electron interacts quite strongly with one nucleus and then somewhat more weakly with those of the surrounding ligand atoms, as is often the case in enzymes containing transition group ions. Two specific examples of these different cases may help to illustrate the kind of hyperfine splittings that are then to be expected.

The case of equal interaction with a number of similar nuclei will be considered first and, since hydrogen atoms are among the atoms most commonly attached to carbon rings or to chains in organic molecules, interaction with a varying number of hydrogen atoms, or protons, will be taken by way of example. It should be noted in this context that the normal isotope of carbon has no such nuclear spin, or magnetic moment, and therefore does not produce a hyperfine pattern; hence, any compound containing only carbon and hydrogen atoms will give a hyperfine pattern which is entirely due to the hydrogen atoms or protons. In a large number of cases in biochemical studies, aromatic compounds are present, and hence the case of a simple benzosemiquinone ring may serve as an illustration of this type of interaction.

The benzosemiquinone free radical can exist in quite a stable form and its basic structure is shown in Fig. 1.13(a); the six-membered carbon ring always has two oxygen atoms attached at opposite ends. In the simple benzosemiquinone radical itself, the other four positions are occupied by protons, but each can be replaced by a chlorine atom and the change to the benzosemiquinone from the tetrachlorinated derivative, in which each of the hydrogen atoms has been replaced by a chlorine atom, will be followed step by step. In Fig. 1.13(a), the four possible forms of the chlorinated benzosemiquinone radical are shown, beginning with the tetrachlorinated version on the left-hand side and moving across to the simple benzosemiquinone radical itself on the right, in which the four chlorines have been replaced by the four protons. The interaction between the unpaired electron, which is moving in the molecular orbital around the ring structure itself, and the atoms on the edge of the ring is then as indicated in Fig. 1.13(b) in the form of an energy-level diagram. The transitions between these energy levels then produce the observed electron resonance spectrum[5] which is shown in Fig. 1.13(c), below the corresponding energy-level system itself. It should be noted that these are first-derivative tracings of the actual absorption and that the chlorine atoms do not interact significantly with the unpaired electron, and therefore no hyperfine structure is produced by these.

Thus, in the first case shown there are no protons present around the aromatic ring, and hence the unpaired electron does not experience any additional incremental magnetic field, and only the two initial electronic energy levels are produced. As a result, a single electron resonance absorption line is obtained as indicated. If, however, one of the chlorine atoms is replaced by a hydrogen atom, the unpaired electron will interact with the single proton, and both the electronic levels will be split into two, since the proton itself can be lined up with or against the direction of the electron spin. The nuclear spin of a proton is $I = \frac{1}{2}$, and thus only the two orientations of $M_I = +\frac{1}{2}$ or $-\frac{1}{2}$ are allowed. This is, therefore, an example of the simple $(2I+1)$ rule noticed earlier and a doublet hyperfine pattern will be obtained as indicated in the figure below.

If *two* of the chlorine atoms are replaced by hydrogen atoms, as shown in the centre of the diagram, the unpaired electron will now interact with these two protons and, as the symmetry of the system shows, its interactions with the two will be equal. Each of the levels, which has already been split into two by interaction with the first proton, will now be split again into two by interaction with the second proton. As a result, three different energy levels are produced in both sets of the electronic states, the centre level in each case being composed of two that have come together as indicated.

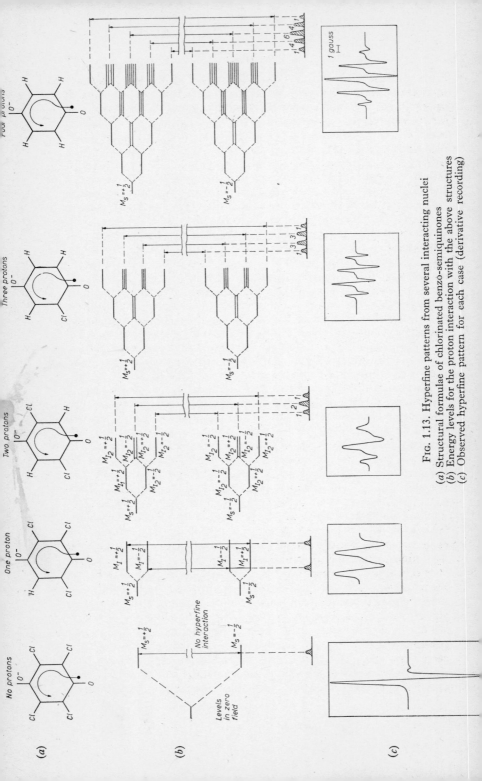

Fig. 1.13. Hyperfine patterns from several interacting nuclei
(a) Structural formulae of chlorinated benzo-semiquinones
(b) Energy levels for the proton interaction with the above structures
(c) Observed hyperfine pattern for each case (derivative recording)

The transitions between these now produce a triplet hyperfine structure, as shown, but with the central line twice as intense as the two outer lines, since it arises from two overlapping sets of levels. Thus an interaction with two equally coupled protons produces a three-line spectrum with an intensity ratio $1 : 2 : 1$. If three of the chlorine atoms are now replaced by hydrogen atoms, the unpaired electron will be interacting with three protons instead of two and a further splitting of the lines will be produced as indicated in the energy-level diagram. The net result is that four different energy levels occur in each electronic state, but the two central ones consist of three components which have come together as shown. The transitions between these thus produce a hyperfine pattern with four lines with intensity ratio $1 : 3 : 3 : 1$. In a similar way, if all four chlorine atoms are replaced by hydrogens, the unpaired electron will then be interacting equally with four protons and the net result is a further splitting of the levels within each electronic state to give five distinct sets, and the transitions between these give rise to hyperfine patterns with an intensity ratio of $1 : 4 : 6 : 4 : 1$, as indicated. It will be evident that this kind of interaction can be extended to any number of equally coupled protons and that, in general, an unpaired electron interacting equally with n protons will produce a hyperfine pattern with $(n + 1)$ lines, the intensity of which varies in a binomial fashion, as already indicated. This variation in intensity across the hyperfine pattern can be immediately identified with an unpaired electron moving in a molecular orbital and coupling to several nuclei. It is quite distinct from the equally intense $(2I + 1)$-line hyperfine pattern which is obtained when the unpaired electron only couples to a single nucleus.

This hyperfine pattern produced when an electron couples with a number of similar nuclei is often found in spectra of organic radicals and will be illustrated in detail in Chapter 4. Another pattern that can also be mentioned here, however, is produced when the orbit of the unpaired electron couples to two nuclei, or to two different sets of nuclei, with interactions differing considerably in strength. This pattern is often met in free-radical spectra as well. Thus n protons may be found equally coupled to the unpaired electron while, in other parts of the molecular orbital, there may be m other protons coupled with a much smaller interaction constant. It will be evident that in this case $(n + 1)$ main lines will be obtained in the spectrum, but that each of them in turn will be split into a further $(m + 1)$ lines by the weak interaction with the other set of protons. This kind of spectrum is again often obtained and will be discussed in more detail later (see for example Fig. 4.1).

1.5.5   *Superhyperfine patterns*

The other case in which two quite different magnitudes of coupling are often obtained is to be found in the study of transition group complexes, and especially as they may arise in enzyme

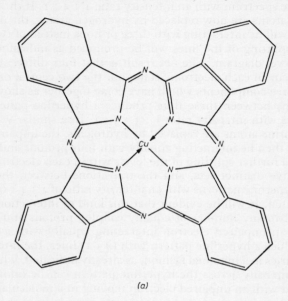

(a)

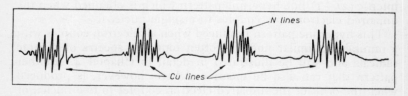

(b)

Fig. 1.14. Superhyperfine pattern observed in copper phthalocyanine

(a) Structural formula of copper phthalocyanine
(b) Observed hyperfine pattern showing the nitrogen superhyperfine splittings on each hyperfine component of the copper spectrum

studies, or the like. The case of an interaction with a copper nucleus may again be taken to illustrate this point. In particular, the case of copper phthalocyanine may be considered, since this molecule, which has the structural form indicated in Fig. 1.14(*a*) has many features in common with several biochemically important molecules. The molecule consists basically of a large flat plane, the copper atom being contained within a square of nitrogen atoms which surround it. These in turn form part of a large conjugated ring system, as shown, and this system is very similar to the porphyrin plane which plays a very important role in a large number of biochemical compounds. The electron resonance spectrum observed from such a copper phthalocyanine molecule is shown in Fig. 1.14(*b*), and four main groups of lines can be clearly seen. These four groups represent the interaction of the unpaired electron with a copper nucleus, as previously discussed, and reflect the $I = \frac{3}{2}$ of the copper nucleus. It is seen, however, that each of these lines also splits into a more complicated superhyperfine pattern and closer inspection reveals that each group of lines consists of two overlapping superhyperfine patterns. The two groups are produced by the two different isotopes of copper, the Cu^{63} and Cu^{65} mentioned previously, and now each produces a slightly shifted set of hyperfine patterns, and each component of these has a further superhyperfine splitting on it.

The superhyperfine pattern present on each of the copper lines can be seen to consist of nine separate individual lines, and these can be attributed to the interaction of the unpaired electron with the four nitrogen atoms surrounding the copper. The nitrogen nucleus has a spin of $I = 1$ and it is evident that if four equally coupled nitrogen atoms are taken together, there are nine different sets of incremental magnetic fields which they could produce. The two extreme cases, in which all the nitrogen nuclei are lined up in one direction or the other, will each, of course, correspond to only one possibility, whereas the intervening situations, in which n of the nitrogen nuclei line up in one direction and $4-n$ in the other, can be obtained in a variety of ways. This accounts for the rising intensity distribution of the superhyperfine pattern towards the centre of the nine lines, in exactly the same manner as was noticed for the protons in the benzosemiquinone structure. The existence and analysis of these superhyperfine patterns will be considered in much greater detail in Chapter 5, where it will be seen that they can give useful additional information in the study of metallo-organic compounds. They have been mentioned here to show that the hyperfine patterns can be used in great detail to map out the actual wave-function of the unpaired electron over either the whole molecule or the particular group or complex with which it is associated.

These few examples have been given by way of illustrating the mechanisms by which the hyperfine splitting itself is produced, and how this can be recognized and analysed in various types of sample. Much more detailed treatment of the whole theory of hyperfine interaction may be found in the introductory volume to this series of monographs,[6] and also in the monograph concerned with transition group atoms, in particular. The main application of these ideas, and the information they give in the study of biochemical and biological systems, is considered itself in much more detail in Chapters 4, 5 and 6.

References

1. Assenheim, H. M., *Introduction to Electron Spin Resonance*, pp. 36–44 (Adam Hilger Ltd, London, 1966; Plenum Press, N.Y., 1966).
2. Ingram, D. J. E., *Spectroscopy at Radio and Microwave Frequencies*, pp. 89–92 and 436–43 (2nd edn, Butterworths, London, 1967).
3. Rubins, R. S., *Electron Spin Resonance in Transition Metal Ions* (proposed for eventual publication by Adam Hilger Ltd, London).
4. Assenheim, H. M., *op. cit.*, pp. 37–40.
5. Wertz, J. E., and Vivo, J. L., *Journ. Chem. Phys.*, **24** (1956), 479.
6. Assenheim, H. M., *op. cit.*, pp. 109–128.

Chapter 2

Experimental Techniques

§2.1

SIMPLE SPECTROMETER SYSTEMS

It will have become evident from the previous chapter that there are, in effect, four basic experimental requirements for the production of electron spin resonance. In the first place a strong homogeneous and stable magnetic field must be available to split the energy levels, as discussed. If work is to be carried out mainly at 3·2 cm wavelength, then this field will be centred around a value of 3000 to 4000 gauss, whereas if the experiments are also to be undertaken at 8 mm wavelength, or shorter, then the magnetic field will need to rise to 13 000 gauss, or above.

In the second place, there must be a suitable source of microwave power, the wavelength and frequency of which is normally chosen from amongst those readily available from a standard radar wavelength band. The third essential requirement is the 'absorption-cell system', i.e. the part of the apparatus which concentrates the incident radiation on the specimen and enables the amount of absorption that takes place within it to be detected. In the case of microwave spectroscopy, this system normally takes the form of a cavity resonator placed between the pole pieces of the electromagnet; and the microwave radiation is fed into it. The fourth and final basic requirement is an electronic system which enables the actual absorption spectra to be displayed, or recorded, in a suitable permanent fashion. The way in which these various items can be put together to form a simple spectrometer is shown in the block diagram of Fig. 2.1.

The microwaves are normally supplied from a radar valve known as a klystron, although solid-state sources, in which harmonics are derived from solid-state devices operating at lower frequencies, have recently begun to be employed. The microwave radiation is fed down a waveguide to the cavity resonator, in which it is effectively concentrated, and in this way a high level of the oscillating microwave magnetic field can be applied to the specimen. A de-

39

tailed consideration of cavity resonators and their use in electron resonance spectrometers may be found in the third volume of this series of monographs.[1] In the simple transmission system shown, a certain amount of power from the cavity resonator is coupled out and fed to a detecting crystal at the far end of the microwave run. The value of the detected current flowing through the crystal is then a measure of the power level existing within the cavity at any given time. Hence, if absorption occurs in the cavity, a reduction will also be produced in the current detected through the crystal. The external magnetic field is applied across the cavity as

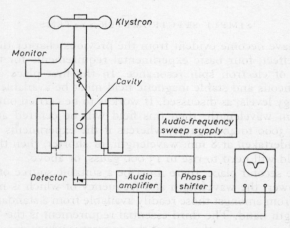

FIG. 2.1. Block diagram of simple transmission-type electron resonance spectrometer
Audio-frequency modulation of the magnetic field produces a 'crystal video' system

shown. This field is normally produced by an electromagnet, since it is the magnetic field and not the frequency that is usually varied to obtain the resonance condition.

In principle, it is possible to obtain the resonance condition either by holding the magnetic field constant and varying the microwave frequency or by holding the microwave frequency constant and varying the magnetic field. The latter method is nearly always adopted, since a large number of the microwave components are frequency sensitive and the microwave system can only be correctly aligned and matched for one given frequency. Once this has been done, it is wise to leave the microwave system set and to interfere with it as little as possible. The microwave oscillators also have a somewhat limited range of frequency coverage. For both these reasons the spectrum is almost always obtained by working with a constant microwave frequency and varying the magnetic field strength. In the simplest type of spectrometer, the field can be changed in small incremental steps; the magnitude of the detected current can be read off at points corresponding to each step and the electron resonance spectrum thus plotted. This is not only time-consuming but also produces very poor sensitivity, so that modulation techniques are always employed which oscillate the magnetic field to and fro through the resonance condition and produce audio-frequency or radio-frequency signals at the detector, whose output can be amplified and displayed electronically.

The simplest form of such modulation is obtained by applying an audio-frequency sweep to the magnetic field with the modulation coils shown, and then applying the same frequency sweep to the time-base of the display oscilloscope, while the amplified signal from the detector is fed to the vertical deflecting plates. In this way the absorption, as it occurs when the field value sweeps through the resonance condition, can be displayed directly on the oscilloscope screen, as shown. This simple type of spectrometer was used for some years in the early resonance work and has associated with it all the basic features of a present-day electron resonance spectrometer. However, as seen in the next section, it does have an inherently low sensitivity owing to the large excess noise which the detecting crystals develop at these low audio frequencies.

§2.2

SENSITIVITY CONSIDERATIONS

Before discussing the design of more refined electron resonance spectrometers, the basic factors which determine the sensitivity of any spectrometer system will first be considered. In studies of the electron resonance spectra of biochemical or biological systems, sensitivity can be one of the most important factors in the experiment. Thus, if the systems are at all close to *in vivo* conditions, they are likely to consist of dilute solutions of the particular compounds under consideration and the number of unpaired electrons present will tend to be rather small. There is also one other effect which tends to reduce the signal strength considerably in biological samples, and that is the presence of water as the solvent. Water possesses a high dielectric constant and also a large damping loss for any oscillating electric field in which it is placed, and hence, if aqueous solutions are to be studied, they must be held in specimen cells which keep the sample well out of the regions of microwave electric field strength. Fig. 2.2 shows the distribution of the magnetic and electric fields as they exist in a simple rectangular cavity, such as is often used in electron resonance spectrometers. It will be seen that the lines of oscillating microwave magnetic field are concentrated vertically down the centre of the cavity, and it is into this region that the specimen

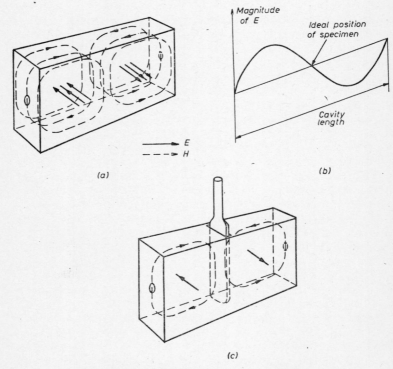

FIG. 2.2. Microwave field patterns in a rectangular cavity
(*a*) Perspective view of electric and magnetic fields
(*b*) Variation of electric field strength through the cavity
(*c*) Sample-holder for aqueous solutions designed to avoid electric field pattern

holder is placed. The electric field has its maximum strength a quarter of a wavelength away from this position on either side, and, in this simple dominant-mode pattern, the electric field takes the form of flux lines perpendicular to the broad waveguide walls. The magnitude of this oscillating electric field dies away from these maximum positions and passes through a zero at the actual point of the maximum magnetic field, as indicated in Fig. 2.2(*b*). It will be seen, however, that the electric field does have some magnitude, even at small distances away from this central spot, and if a highly lossy medium, such as an aqueous solution, is present, its volume must be restricted to a region very close to the central null portion. Thus spectrometers for the study of aqueous solutions are normally provided with a special quartz specimen-holder of rectangular shape, very narrow along the direction of the waveguide itself, but spread out to fill the whole width of the guide, as indicated in Fig. 2.2(*c*). In this way a maximum amount of sample can be inserted into the oscillating microwave field without penetrating into the region of electric field. The 'filling factor', which measures the amount of sample in the microwave field compared with the total volume occupied by the microwave field, is nevertheless noticeably smaller than the filling factor which can be obtained with a non-lossy specimen, and this fact tends to reduce the sensitivity that can be achieved in a large number of biological studies. For these reasons, it is important when studying aqueous solutions to design spectrometer systems which are as sensitive as possible.

The other basic parameter which defines the quality and usefulness of a spectrometer system is its resolution, or, in other words, the minimum separation between absorption lines that can be distinguished in the spectrum. The resolution obtainable is sometimes determined by the specimen itself, and any substance which gives rise to wide absorption lines will obviously prevent fine detail from being observed in the spectrum. On the other hand, if solutions are being studied, quite a number of the broadening processes present are often averaged to zero, as mentioned previously, and it is then possible to observe very narrow absorption lines in electron resonance spectra. In such cases as these, the resolution is often limited by the instrumentation of the spectrometer itself, the most likely source of limitation being the imperfect uniformity or homogeneity of the magnetic field. If superhyperfine components are to be observed with a separation of 30 milligauss or less, the homogeneity of the magnetic field over the volume of the specimen must be better than 30 milligauss or the individual lines will never be resolved. This homogeneity of the magnetic field must also be achieved in time variation, as well as in spatial variation over the volume of the specimen. Thus, if it takes 20 seconds to trace through the spectrum and the field itself is unstable, changing by more than 30 milligauss during the time of the tracing, the spectrum will lose its resolution and be smeared out. Exactly the same reasoning clearly applies to the stability of the microwave frequency; if frequency changes during the recording of the spectrum, the fine splittings will again be smeared out and lost. As a general comment, therefore, it can be said that the resolution of any electron resonance system depends basically on the time stability of the microwave frequency and magnetic field and also on the spatial uniformity of the magnetic field over the specimen. These factors are considered in detail in Wilmshurst's book[2] and will not be treated any further here.

Returning now to a determination of the actual parameters which define the sensitivity of the spectrometer system, or the minimum number of unpaired electrons which are capable of detection, the following procedure can be adopted. The basic limitation on sensitivity will be due to the random noise produced within the detecting system itself, and the one element which produces much more noise than any other is the semiconductor crystal detector at the end of the waveguide run. A calculation of the ultimate sensitivity of the spectrometer can therefore be made by determining the noise voltage produced across this detecting crystal and then calculating the susceptibility of the sample, or the number of electrons present in it, which would produce the same value of voltage as an absorption signal. If the detecting crystal were a perfect detector and introduced no excess noise of its own, the noise voltage developed across it could be written as

$$V_{noise} = (R_0 \, kT. \, \Delta v)^{\frac{1}{2}} \qquad (2.1)$$

where R_0 is the waveguide impedance, k is Boltzmann's constant, and T is the absolute temperature.

In practice, however, noise from quite a number of additional sources may be included in a general expression giving the noise power developed at the output of the detector, thus

$$P_{noise} = \left(\frac{N_m}{L} + F_A + t_D - 1 \right) kT. \, \Delta v \qquad (2.2)$$

where N_m is the incident noise from the microwave run, L is the conversion loss of the crystal, F_A is the noise of the amplifier, and t_D is the noise temperature of the crystal itself.

In order to calculate what susceptibility, or unpaired electron concentration, would produce a signal across the detector of this same magnitude, the effect of the absorption in the cavity must first be considered. The absorption of the microwave power is due to the interaction of the magnetic moments of the unpaired electrons with the magnetic field of the microwave oscillations. The unpaired electrons have an effective susceptibility given by

$$\chi = \chi' - j \cdot \chi'' \qquad (2.3)$$

where χ' represents the real or 'in-phase' susceptibility, and χ'' the imaginary, or 'out-of-phase', susceptibility. Thus, if an oscillating magnetic field, given by $H_1 \sin \omega t$, is applied across the specimen, a magnetization will be induced with the following components:

$$\chi' H_1 \sin \omega t \quad \text{(in-phase magnetization)}$$

and $\quad - \chi'' H_1 \cos \omega t \quad \text{(out-of-phase magnetization)}$

The general expression for the absorption of energy in a magnetic interaction is $H(dM/dt)$; hence, the total power absorbed by the unpaired electrons in the specimen can be obtained by dividing the total energy absorbed during one complete cycle by the duration of the cycle. Thus

$$\text{Power absorbed} = P_a = \frac{1}{2\pi/\omega} \int_{\text{cycle}} H \frac{dM}{dt} \, dt \qquad (2.4)$$

Here the $2\pi/\omega$ is the time of one cycle, and the integral is also taken over this period. Thus

$$P_a = \frac{\omega}{2\pi} \int_0^{2\pi/\omega} H \frac{dM}{dt} \, dt$$

Therefore

$$P_a = \frac{\omega}{2\pi} H_1{}^2 \int_0^{2\pi/\omega} \omega \sin \omega t \, [\chi' \cos \omega t + \chi'' \sin \omega t] \, dt$$

Therefore

$$P_a = \tfrac{1}{2} . \omega . \chi'' . H_1{}^2 \qquad (2.5)$$

The effect of this absorption on the Q-factor of the cavity must now be considered. This Q-factor measures the energy stored in the cavity compared with the energy dissipated by resistive wall losses, coupling holes, or any other causes. Thus, if no electron resonance absorption is taking place, the Q-factor can be defined as

$$Q_0 = \omega . \frac{\text{Energy stored}}{\text{Power dissipated}}$$

Therefore $\qquad Q_0 = \dfrac{\dfrac{\omega}{8\pi} \displaystyle\int_{V_c} H_1{}^2 \, dV_c}{P_w} \qquad (2.6)$

where integration over the cavity volume, V_c, of the microwave magnetic field gives the energy stored within the cavity, and the P_w of the denominator represents the power that is dissipated owing to the resistance of the cavity walls or to leakage through the coupling holes.

If electron resonance absorption now occurs there will be an additional form of power dissipated and therefore the Q of the cavity will be reduced and its new value will be given by the expression

$$Q_{\text{absorption}} = \frac{\dfrac{\omega}{8\pi} \displaystyle\int_{V_c} H_1^2 \, dV_c}{P_w + \tfrac{1}{2}\omega \cdot \chi'' \cdot \displaystyle\int_{V_s} H_1^2 \, dV_s} \qquad (2.7)$$

In this case, a new integral is added to the denominator and this represents the interaction of the microwave magnetic field with the unpaired electrons in the sample, as discussed above and given in equation (2.5). It should be noted that in this case the integral is taken over the volume of the specimen, V_s, and not over the whole volume of the cavity, V_c.

It can be assumed that the power absorbed in the specimen is small compared with the other power losses, and then the denominator can be divided by P_w and expanded to give

$$Q_{\text{absorption}} = \frac{\dfrac{\omega}{8\pi} \displaystyle\int_{V_c} H_1^2 \, dV_c}{P_w} \left(1 - \frac{1}{P_w} \cdot \frac{1}{2} \cdot \omega \cdot \chi'' \int_{V_s} H_1^2 \, dV_s + \dots \right)$$

$$(2.8)$$

Therefore

$$Q_{\text{absorption}} = Q_0 \left(1 - \tfrac{1}{2}\omega \cdot \chi'' \cdot \int_{V_s} H_1^2 \, dV_s \cdot Q_0 \Big/ \frac{\omega}{8\pi} \int_{V_c} H_1^2 \cdot dV_c \right)$$

Therefore

$$Q_{\text{absorption}} = Q_0 \left[1 - 4\pi \cdot \eta \cdot \chi'' \cdot Q_0 \right] \qquad (2.9)$$

It may be noted that the filling factor, η, has been substituted for the ratio of the two integrals of the microwave magnetic field distribution and that it measures how much of the effective microwave magnetic field is in fact filled by the specimen.

Hence the change, ΔQ, produced by the absorption due to the unpaired electrons is given by

$$\Delta Q = Q_0{}^2 \cdot 4\pi \cdot \eta \cdot \chi'' \tag{2.10}$$

and the relative change in Q, which can be equated to a relative change in an equivalent damping resistance, R, is given by

$$\frac{\Delta Q}{Q} = \frac{\Delta R}{R} = 4\pi \cdot \eta \cdot \chi'' \cdot Q_0 \tag{2.11}$$

In order to calculate the change which this change in the Q-factor of the cavity produces in the voltage across the detector, a low-frequency equivalent circuit of the microwave system is required. Such an equivalent circuit for a simple transmission-type E.S.R. spectrometer is shown in Fig. 2.3, where the klystron

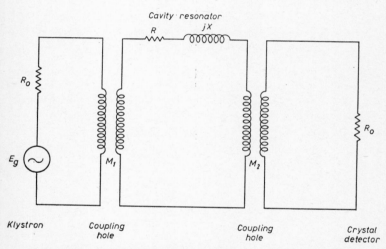

FIG. 2.3. Equivalent circuit for transmission-type E.S.R. spectrometer
The klystron is represented by the generator E_g and the cavity by the central circuit with resistive loss R and reactance X

valve, acting as the microwave power source, is represented by the generator E_g with an impedance R_0. This couples via the cavity coupling holes, represented by the mutual inductance M_1, into the cavity, represented by the central circuit with resistive losses, R, and reactance, X. The power is then coupled out via the second coupling hole, represented by M_2, to the crystal detector, represented by the matched resistor, R_0, at the end of the circuit and across which the detected voltage V_L is developed.

The magnitude of this detector voltage can then be derived in terms of the other circuit constants[3] and expressed as

$$V_L = E_g \cdot \frac{\omega^2 M_1 \dfrac{M_2}{R_0}}{R + jX + \dfrac{\omega^2 M_1{}^2}{R_0} + \dfrac{\omega^2 M_2{}^2}{R_0}} \qquad (2.12)$$

The change produced in this detector voltage by the absorption by the unpaired electrons can then be calculated as the change in V_L which corresponds to a small change in the resistive loss of the circuit, R, thus

$$\frac{\delta V_L}{\delta R} = -E_g \cdot \frac{\dfrac{\omega^2 M_1 M_2}{R_0}}{\left(R + jX + \dfrac{\omega^2 M_1{}^2}{R_0} + \dfrac{\omega^2 M_2{}^2}{R_0} \right)^2} \qquad (2.13)$$

The maximum sensitivity will be obtained for the spectrometer when this expression for $\delta V_L / \delta R$ is maximized. It can be seen that this implies immediately that $X = 0$ (i.e. that the cavity should be tuned to resonance), and optimization with respect to the coupling parameters gives

$$\frac{\omega^2 M_1{}^2}{R_0{}^2} = \frac{\omega^2 M_2{}^2}{R_0{}^2} = \frac{R}{2R_0} \qquad (2.14)$$

Thus the maximum value of $\delta V_L / \delta R$ will be given by

$$\frac{\delta V_L}{\delta R} = -\frac{E_g}{8R} \qquad (2.15)$$

and hence the actual change in voltage developed across the detector in terms of the change in Q of the cavity will be

$$\delta V_L = -\frac{E_g}{8} \cdot \frac{\delta R}{R} = -\frac{E_g}{8} \cdot \frac{\Delta Q}{Q_0} \qquad (2.16)$$

Therefore, from equation (2.11),

$$\delta V_L = -\frac{E_g}{8} 4\pi \cdot \eta \cdot \chi'' \cdot Q_0$$

Therefore $$\delta V_L = -\tfrac{1}{2} E_g \cdot \pi \cdot \eta \cdot \chi'' \cdot Q_0 \qquad (2.17)$$

The minimum detectable susceptibility will therefore be obtained by equating the value of this change in voltage to the noise voltage developed across the detector and given by equation (2.1). Thus

$$\tfrac{1}{2} E_g . \pi . \eta . \chi'' . Q_0 = (R_0\, kT . \Delta v)^{\tfrac{1}{2}}$$

Therefore $\qquad \chi''_{min} = \dfrac{1}{\pi . \eta . Q_0} \left(\dfrac{4\, R_0 . kT . \Delta v}{E_g{}^2} \right)^{\tfrac{1}{2}}$

Therefore $\qquad \chi''_{min} = \dfrac{1}{\pi . \eta . Q_0} \left(\dfrac{kT . \Delta v}{P_0} \right)^{\tfrac{1}{2}} \qquad (2.18)$

where P_0 has been written in place of $E_g{}^2/4R_0$ and represents the maximum available microwave power from the klystron valve.

It should be noted that this expression has been deduced from equation (2.1) which assumed that the crystal detector had no excess noise. As already briefly noted in equation (2.2), however, there are various additional sources of noise which can all be included in a noise figure, F, where

$$F = \left(\frac{N_m}{L} + F_A + t_D - 1 \right) \qquad (2.19)$$

as defined in equation (2.2); and the practical expression for the minimum detectable susceptibility then becomes

$$\chi''_{min} = \frac{1}{\pi . \eta . Q_0} \left(\frac{F . kT . \Delta v}{P_0} \right)^{\tfrac{1}{2}} \qquad (2.20)$$

The property of the specimen of practical interest, especially in biological and biochemical applications, is the number of unpaired electrons it contains, rather than the value of the complex susceptibility as given in equation (2.20). The final step in the calculation of the practical sensitivity of the electron resonance spectrometer is, therefore, to find a relation between the number of unpaired electrons present in the specimen and the magnitude of the complex susceptibility which they produce. This may be approached by first considering the relation between the number of unpaired electrons and the d.c. susceptibility, as is derived by ordinary paramagnetic theory.

The magnetization, and thus also the susceptibility of any specimen, will be determined by the difference in population between the spins with moments aligned along the field and those with spins aligned against the field. It has already been seen in equation (1.3) that the ratio of these two populations is given by $e^{-hv/kT}$, where v is the frequency of the microwave radiation being applied. This energy hv could be replaced equally well by the expression $g\beta H$. Under all normal conditions, this energy is much smaller than that associated with the temperature of the specimen and equal to kT, and so $g\beta H/kT$ is small and the exponential can be approximated by expanding it in a power series and neglecting all but the first two terms to give $(1 - g\beta H/kT)$. This approximation may not be valid at liquid helium temperatures and below, but at liquid hydrogen temperatures of $20°$ K and above it will certainly hold good, and hence in nearly all experiments associated with biological or biochemical specimens it can be taken as quite valid. It will be seen from this that the net difference in population between the two groups of electrons can now be put equal to $(N_0/2)(g\beta H/kT)$, where N_0 is the total number of unpaired electrons and thus the number of reactive atoms or molecules.

Each of the excess electrons will have a resolved magnetic moment of $\frac{1}{2}g\beta$ along the direction of the applied field, and there will thus be a net magnetic moment which is not cancelled out by other electrons and is equal to $(g^2.\beta^2.N_0.H)/(4kT)$. It follows that the d.c. mass susceptibility of the specimen can be written as this quantity divided by the magnetic field producing it, or in other words that

$$\chi_0 = \frac{1}{4kT}g^2.\beta^2.N_0 \qquad (2.21)$$

Numerical values can be substituted for the atomic constants in this equation, and if it is assumed that the g-value is very close to that of the free spin, as will be so for any free-radical-type specimen, the actual relation between N_0, the total number of the unpaired electrons in the specimen, and the d.c. susceptibility can be written as

$$N_0 = 1·60.10^{24}.T.\chi_0 \qquad (2.22)$$

where χ_0 is quoted in c.g.s. electromagnetic units.

An expression relating the number of unpaired electrons in the specimen to the d.c. susceptibility has thus been obtained, but it should be noted that in this case we have considered only electrons distributed between two possible energy levels, and thus having only one unit of spin angular momentum $S = \frac{1}{2}$. This will normally be true in all biological and biochemical specimens, except for those which contain transition group ions, where there may be more than one unpaired spin per atom. In such cases, the general theory of paramagnetic susceptibility can be very easily modified, since the population distribution between the $(2S+1)$ different levels produced by the magnetic field can be calculated with the same approximation as before. The magnetization produced by each of these levels can be summed to give the net results for all possible resolved components of the magnetic moments. The more general expression for the d.c. mass susceptibility in terms of the number of unpaired electrons then becomes[4]

$$\chi_0 = \frac{1}{3kT} g^2 . \beta^2 . S(S+1) . N_0 \qquad (2.23)$$

For the rest of this section it will be assumed, however, that there is only one unpaired electron per atom or molecule in the specimen being considered, and hence equations (2.21) and (2.22) will be used, although the argument that follows can equally well be applied to the more general case represented by equation (2.23) if required.

The final step in this calculation will therefore be to obtain a relation between the d.c. susceptibility χ_0 and the complex susceptibility χ''. It will be remembered that the idea of the complex susceptibility was initially introduced when the amount of microwave power absorbed was being considered. This power absorbed can also be derived by considering the actual number of electronic transitions between the two states in question. Thus for the simple case of one unpaired electron per atom or molecule, the difference in population between the two levels has been shown to be

$$\frac{N_0}{2} . \frac{g\beta H}{kT} \quad \text{or} \quad \frac{N_0}{2} . \frac{\hbar\omega}{kT}$$

and each of the transitions between these levels will produce absorption of energy $\hbar\omega$, where ω is the angular frequency. The resultant net absorption of microwave power due to transitions between these two levels will then be given by

$$P = \frac{N_0}{2} . \frac{\hbar^2 . \omega^2}{kT} . p_m \qquad (2.24)$$

where the value of p_m is given by standard radiation theory as

$$p_m = \frac{\pi}{4} . \gamma^2 . H_1^2 . g(\omega - \omega_0) \qquad (2.25)$$

In this equation, γ is the gyromagnetic ratio of the electron and $g(\omega - \omega_0)$ is the line-shape function as already defined in equation (1.6). The power absorbed by the transitions between the two levels is therefore given by combining equations (2.24) and (2.25) to give

$$P_{abs} = \frac{\pi}{8} . \frac{g^2 . \beta^2 . H_1^2 . N_0}{\hbar^2 \, kT} . \hbar^2 . \omega^2 . g(\omega - \omega_0) \qquad (2.26)$$

In this equation, the gyromagnetic ratio γ has been put equal to $g\beta/\hbar$ and the expression thus obtained can now be related to the expression for the d.c. susceptibility given in equation (2.21) and hence the power absorbed written as

$$P_{abs} = \frac{\pi}{2} H_1{}^2 . \omega^2 . \chi_0 . g(\omega - \omega_0) \qquad (2.27)$$

This expression for the absorbed power may now be equated to the expression derived earlier when the idea of a complex susceptibility was first introduced, as given in equation (2.5); and the relation (2.28) is thus obtained.

$$\chi'' = \chi_0 . \pi . \omega . g(\omega - \omega_0) \qquad (2.28)$$

It will be remembered, when discussing line-shape function earlier in §1.3.2 that a parameter T_2 was defined by equation (1.9) and it was seen that the inverse of this parameter was a measure of the width of the absorption line. The line-width parameter, $\Delta\omega$, can therefore be best described as $1/\pi . g_{max}$, where g_{max} is the maximum value of the line-shape function $g(\omega - \omega_0)$ as obtained at the resonance frequency ω_0. It follows that, at the resonance frequency itself, equation (2.28) becomes

$$\chi'' = \chi_0 \left(\frac{\omega_0}{\Delta\omega} \right) \qquad (2.29)$$

and that this may in turn be combined with equation (2.21) to give a final expression for the relation between the complex susceptibility and the number of unpaired electrons in the sample as

$$N_0 = \left(\frac{4kT}{g^2 . \beta^2} \right) . \left(\frac{\Delta\omega}{\omega_0} \right) . \chi'' \qquad (2.30)$$

This equation may now finally be taken together with equation (2.20), which expressed the minimum detectable complex susceptibility in terms of the actual parameters of the electron resonance spectrometer, to give the final equation

$$N_{0min} = \underbrace{\left(\frac{4k}{g^2 . \beta^2} \right) T_{spec}}_{\substack{\text{Specimen}}} \times \underbrace{\frac{\Delta\omega}{\omega_0}}_{\substack{\text{Resonance} \\ \text{condition}}} \times \underbrace{\frac{1}{Q_0 . \pi . \eta}}_{\substack{\text{Cavity}}} \times \underbrace{\left(\frac{F . kT_{Det} . \Delta v}{P_0} \right)^{\frac{1}{2}}}_{\substack{\text{Detecting} \\ \text{system}}}$$

$$(2.31)$$

This equation gives a definite relation between the minimum number of unpaired electrons that can be detected in a specimen and the practical parameters associated with the whole of the electron resonance spectrometer system. These different parameters have been divided into four main groups, as indicated below the equation, the first group containing just the fundamental atomic constants and the temperature of the specimen. It should be noted that this temperature is normally quite different from that of the detector, which comes in the fourth group at the end of the expression. The second group of parameters consists of the resonance frequency, ω_0, and the line-width parameter, and so is partially determined by the experimental conditions, governed by ω_0, and partially by the properties of the specimen itself as defined by $\Delta\omega$. The third group of parameters is entirely concerned with the properties of the resonance cavity into which the specimen is placed, and in particular the Q-factor of a cavity and the filling factor of the specimen that is placed within it. The fourth group of parameters is entirely concerned with the properties of the detecting system and the available power that is passing through the spectrometer. It is now possible to consider the effect of each of these groups of parameters on the ultimate sensitivity of the spectrometer and to see how this can be optimized in any given case.

The various ways in which the sensitivity can be improved by optimization of these different parameters can be briefly listed as follows.

(a) Reduction of the temperature of the specimen

It is evident that this is the only parameter in the first group of equation (2.31) that can be altered in any way, and if the approximation assumed at the beginning of the calculation is true, a linear increase in sensitivity will be obtained as the temperature is reduced. There may also be other reasons for reducing the temperature, such as to decrease the line-width parameter, $\Delta\omega$. On the other hand, there may be overriding requirements that the experiments be carried out at room temperature, and this applies especially to biological and biochemical systems, where the actual conditions of the study should be as close to *in vivo* conditions as possible. The increase of sensitivity by reduction of the temperature of the specimen may therefore be ruled out in many such applications.

(b) Increase of the resonance frequency

Again if the assumptions of the calculation are true, it follows that the sensitivity of the spectrometer will increase directly with the value of the resonance frequency, although other parameters such as the increasing value of the Q-factor at higher frequencies may, in fact, produce a more rapid increase with frequency than this. It follows, however, that measurements at higher frequencies should always give higher absolute sensitivities, and this is one of the reasons why it is often better to study single crystals at 8-mm wavelength (Q band) rather than 3-cm wavelength (X band).

As mentioned earlier, the 13 000 gauss magnetic field associated with the 8-mm-wavelength spectrometers has previously given an upper limit to the frequency of operation. The advent of super-conducting magnets has materially altered this situation, however, and a considerable amount of work is now proceeding at wavelengths of 4 mm, and even at 2 mm, where the sensitivity in studying the small crystals normally associated with the biochemical specimens can be very noticeably improved. Details of these very high-frequency millimetre spectrometers are given in §3.6, and so further consideration of them will be left until then.

(c) Reduction of the absorption line width

It is fairly clear intuitively, as well as directly from equation (2.31), that a narrow sharp absorption line will be easier to detect than a broad line of the same integrated intensity, and a direct relation between sensitivity and width of the absorption line is to be expected. This width is normally controlled by the conditions within the specimen itself, however, and hence it is normally not a parameter that is open to optimization, although in some cases cooling the specimen can reduce $\Delta\omega$ and thus improve the sensitivity in this way as well.

(d) Increase of cavity Q-factor

An increase in the Q-factor of the cavity will give a direct increase in the sensitivity as indicated, and it is for this reason that cavities are made to limits which are as precise as possible and with surfaces as highly polished as possible. Q-factors can also be increased somewhat by using silver or gold plating, although it must be remembered that the insertion of the specimen itself, even when off resonance, produces an effective Q which is noticeably lower than that of an empty cavity. Careful design of resonators with high Q-factors is, however, an important feature of electron resonance spectrometers.

(e) Increase of the filling factor

The filling factor effectively measures the efficiency of placing the specimen within the total microwave magnetic field available in the cavity; and a high filling factor demands careful geometrical design of the specimen-holder. A compromise often has to be made, however, with the damping, due to electrical losses, that occurs if the sample extends significantly into the region of the electric microwave field. This point has already been mentioned for the case of aqueous solutions and illustrated in detail in Fig. 2.2. Hence, filling factors are often limited by the maximum size of specimen-holder consistent with the damping effect. It should also be noted that high filling factors often reduce the Q-factor appreciably and that the real parameter to optimize is the product $Q \times \eta$.

(f) Increase of microwave power level

The sensitivity of the spectrometer will rise as the square root of the available power level, provided such features as saturation are not present in the specimen. The power level is determined entirely by the availability of klystron valves with high power outputs, and is a point to watch when purchasing or constructing a spectrometer. However, it must be noted that, should the microwave power rise too high, many specimens of interest in the biological and biochemical field will suffer saturation broadening, so that sensitivity cannot be increased indefinitely by increasing this particular parameter. A full consideration of the effects of saturation broadening is given in §3.7.

(g) Decreasing the detector noise level

The different factors which govern the excess noise produced in a detecting system have already been mentioned briefly and summarized in equation (2.2). They are also discussed in detail in the next section, since they are the main practical limitation on the spectrometer sensitivity. The various methods of reducing the different forms of excess noise are also considered in detail there.

(h) Decreasing the bandwidth of the detecting system

It can be seen from equation (2.31) that in principle the sensitivity of the spectrometer can be increased indefinitely by continuously reducing the bandwidth, Δv, of the detecting system. Such a decrease in bandwidth implies an increase of the time constant associated with the spectrometer, and hence also high stability in all of its parts over a long time. This is the ultimate limitation on any such spectrometer system and explains why such care has to be taken in stabilizing both the magnetic field and the microwave frequency. If such highly stable conditions can be provided, however, it is possible to increase the sensitivity significantly by the use of long recording times and circuits with a very narrow

E

bandwidth. Moreover, this long time constant can be obtained by summing over smaller individual time sweeps, which is often more convenient in practice. The whole technique of averaging and integrating methods is discussed in detail later in §3.2, since it is one of the more important recent developments in experimental electron resonance spectrometry.

§2.3

DETECTOR NOISE AND ITS REDUCTION

2.3.1   *Factors affecting spectrometer sensitivity*

As seen in the previous sections, the noise associated with the microwave crystal detector is one of the most crucial parameters in determining the ultimate sensitivity of the whole spectrometer system. It is at this point that the actual microwave signal is converted to the lower frequency modulation, which actually represents the absorption by the specimen, and any excess noise that is introduced at this stage will always be amplified and reproduced by later stages. It is thus imperative to keep the noise associated with the crystal detector to a minimum, and it is this requirement which effectively dominates the design of every electron resonance spectrometer. In the first simple analysis of the transmission-type spectrometer, it was assumed that the noise associated with the detector was entirely due to random thermal fluctuations, and thus could be represented by equation (2.1). In fact, the noise associated with a crystal detector, placed at the end of the waveguide run of a simple spectrometer system, as shown in Fig. 2.1, would be very much greater than the basic thermal noise for two main reasons:

(*a*) The excess noise associated with the crystal increases markedly with the value of the d.c. current flowing through the crystal, and in the system shown this is likely to be quite large.

(*b*) The excess noise increases with the inverse of the frequency of modulation; thus a low-frequency modulation and detection, as represented by the crystal-video system of Fig. 2.1, would also produce large excess low-frequency flicker noise of this type. It therefore follows that all the parameters which introduce excess noise into the crystal must be carefully studied and the optimum conditions chosen for each of them if high sensitivity is to be obtained.

2.3.2   *Noise variation with crystal current*

The way in which the excess noise associated with the crystal varies with the mean value of the d.c. current flowing through it is shown in Fig. 2.4, and it is seen there that the effective noise temperature rises more or less linearly with an increasing mean current. This fact by itself would suggest that it is always best to work with values of mean current in the crystal as small as possible; but another parameter associated with the crystal must be considered at the same time—the conversion loss, which measures the reduction in detection efficiency. This parameter, normally denoted by L, is also plotted on Fig. 2.4; it has a value which increases very rapidly as the crystal current drops to small magnitudes, but falls to a low constant value as the crystal current increases. Some

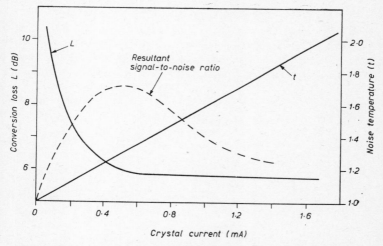

FIG. 2.4. Variation of conversion loss and excess noise of crystal detector with detected current

Variation of the conversion loss is indicated by the curve labelled L and that of the excess crystal noise by the linear variation t. The resultant signal-to-noise ratio is then indicated by the broken curve

compromise must, therefore, be made between these two factors, since it is no use having a low excess noise if the signal itself is detected inefficiently. The effect of combining these two parameters is shown by the broken curve in Fig. 2.4, and it will be clear from this that there is an optimum mean d.c. crystal current of about $\frac{1}{2}$ mA, at which the loss has fallen to a reasonable value and the excess noise has not risen too much. The correct value of this optimum mean current will vary somewhat from crystal to crystal, and is best found in each case empirically. It is evident, however, that some means must be provided in the spectrometer system whereby the mean level of the detecting current through the crystal can be adjusted to a magnitude of about $\frac{1}{2}$ mA.

It is quite clear that the simple transmission system of Fig. 2.1 does not possess such an independent adjustment of the crystal current. The magnitude of the power flowing through the cavity resonator, and on to the crystal detector, will be determined by the optimum coupling of the holes feeding the microwaves into

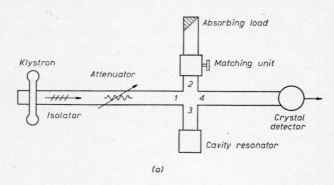

(a)

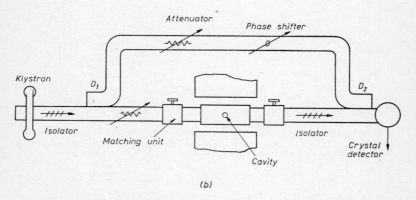

(b)

FIG. 2.5. Microwave balancing circuits as used in E.S.R. spectrometers
(a) Four-arm bridge employing Magic T
(b) Microwave bucking system employing directional couplers (D_1 and D_2)

and out of the cavity resonator, as defined by equation (2.14). The microwave power at the specimen itself is generally adjusted to as high a value as possible, and thus the coupling holes produce quite a large level of power at the crystal detector, and thus a large mean detected current. The crystal will, therefore, have a large excess noise associated with it, and hence will produce a poor sensitivity. There are two basic ways in which independent adjustment can be obtained for the power level within the cavity itself and the power level falling on the crystal detector; the former normally needs to be as high as possible to increase the size of the actual absorption signal, while the latter needs to be optimized to give the $\frac{1}{2}$ mA in the crystal. These two methods are indicated in Fig. 2.5. In the first case, as illustrated in Fig. 2.5(a), a microwave bridge system is used to balance the cavity containing the specimen to be studied, with a matched load in an equivalent arm. The four arms of this microwave bridge are labelled 1, 2, 3, and 4; they correspond to the four arms of any low-frequency a.c. bridge system. The microwave circuits can be designed so that the bridge possesses exactly the same properties as its lower-frequency analogue. Thus if the impedances of arms 2 and 3 are correctly balanced against each other, it can be arranged that the power fed into arm 1 is equally distributed between arms 2 and 3 and that no power is fed to arm 4. Any unbalance of the bridge, produced by a change in the impedance of arm 3, will then result in an unbalance of the null condition, and hence power will be detected in arm 4. The name given to the double T-junction, produced by four such waveguides, is a 'magic-T'. Such a junction has to be carefully designed, but details will not be given here; references can be found in other texts.[5]

It is evident, however, that the detecting crystal can now be mounted at the end of arm 4. If a microwave matching element is included in arm 2, the bridge can be initially balanced so that no power falls on the crystal, and hence no mean detected current is obtained. The condition for optimum sensitivity, however, is not zero detected current, but rather a mean current of about $\frac{1}{2}$ mA. In practice, therefore, the bridge is balanced first to an accurate null point, and then unbalanced slightly so that a mean detected crystal current of about $\frac{1}{2}$ mA is obtained. Any further small unbalance of the bridge, such as is produced by the electron resonance absorption itself in arm 3, is then reflected as a corresponding change in the power fed into arm 4, and hence there is a direct change in the output of the crystal detector. It is thus possible to have high power levels within the cavity itself, with corresponding large values of the actual absorption signal, and yet, at the same time, reduce to any required value the power falling on the detecting crystal. A large number of modern electron resonance spectrometers do therefore employ these magic-T units as the central microwave element distributing the power in the way indicated.

There is one other way, however, in which independent adjustment of power in the cavity and that on the crystal can also be

obtained, and which has somewhat more general application. This
is known as 'microwave bucking', and is illustrated in Fig. 2.5(*b*).
This microwave circuit can be regarded as a simple transmission
system, as was described in Fig. 2.1, but with a bypass arm added
to it, which extracts some of the microwave power via directional
coupler, D_1, and feeds it into a separate waveguide run. This run
contains an attenuator and phase shifter, as shown. The power is
fed back into the main waveguide run, after the cavity resonator
and just before the crystal detector, by the directional coupler D_2,

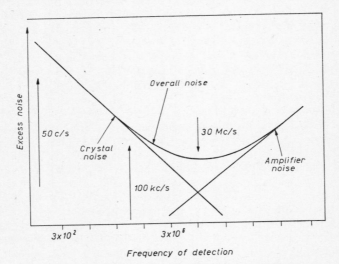

FIG. 2.6. Variation of excess crystal noise with frequency
The linear variation of crystal noise with the inverse of the frequency
is shown on the left-hand side, while a typical rising curve, due to the
intermediate frequency amplifier, is shown on the right-hand side

as shown in Fig. 2.5(*b*). The phase of the microwave radiation fed
back into the main run can now be adjusted by the phase-shifter,
and hence it either adds to, or subtracts from, the radiation which
has been passing down the main waveguide run. Its magnitude can
also be adjusted by the attenuator, and thus these two controls can
produce destructive, or constructive, interference of any desired
magnitude between the two sets of microwave power. Thus the
actual power level falling on the crystal at the end of the run can be
altered continuously and quite independently of the actual power
flowing through the cavity resonator itself. The absorption signal,
which is produced by the specimen in the cavity, will only affect

one of the two microwave components, and hence it is in no way balanced out, and is passed on to the detecting crystal, as a constant value of modulation, on a variable magnitude of carrier frequency.

The advantage of this bucking technique is that it can be applied to any form of spectrometer system. It is now being increasingly used in spectrometers which employ microwave circulators in place of the magic-T's described above. A microwave circulator is a device which makes use of the non-reciprocal transmission properties associated with magnetized ferrite specimens, which can be placed in the waveguide run. Such a circulator is shown in use in Fig. 3.6, which represents the block diagram of a modern spectrometer using such a circulator with a microwave bucking system. The power from the klystron entering in arm 1 is now all fed down to the cavity at the end of arm 2, and then the reflected power from the cavity is all fed on to arm 3 of the circulator, and hence, after mixing with the power from the bucking arm, proceeds to the crystal detector at the end of the run. One great advantage such a microwave circulator has over the magic-T is that all the available power from the klystron is passed into the cavity and none is lost in a balancing arm. Full details of the spectrometer systems employing such circulators may be found in the texts by Ingram and Poole[6] and a detailed discussion of them will not be given here.

2.3.3 *Variation of crystal noise with frequency of detection*

The other parameter which markedly affects the excess noise of the crystal detector is the frequency of modulation, or detection, and the way in which the excess noise varies with this frequency is shown in Fig. 2.6. The excess noise falls linearly with increasing frequency of detection and reaches negligible values at frequencies above about 50 Mc/s. Another factor comes into play however at higher frequencies, and this is the noise of the first-stage amplifiers, which follow the crystal detector itself. These can be designed to have very low noise figures at the lower frequencies, but above 10 Mc/s they will start to have significant additional noise which rises with frequency as shown. The overall noise figure thus has the form indicated by the central curve, and can be seen to have a broad optimum centred at about 30 Mc/s. Hence high sensitivity in microwave detection systems will only be achieved if the detecting, or intermediate, frequency is of about this value, and it is for this reason that the intermediate frequencies used in normal radar sets are at either 30 or 60 Mc/s. It therefore follows that high-sensitivity electron resonance spectrometers must be designed to have modulation frequencies much higher than the audio-frequencies shown in the simple spectrometer of Fig. 2.1;

they should have frequencies of detection within the broad optimum indicated in Fig. 2.6.

The simplest way in which such frequencies can be introduced into the spectrometer is to modulate the magnetic field itself at the higher frequency, instead of at the audio-frequency proposed earlier. Such spectrometers are then called 'high-frequency field-modulation spectrometers' and employ basically the same microwave circuits discussed in §2.3.2 above, but with the additional feature of a high-frequency modulation applied to the magnetic field itself. There is, however, one fundamental drawback to the application of such a simple high-frequency field modulation, and this is the additional broadening which modulation of the actual resonance condition can produce. Just as frequency modulation of an ordinary carrier wave produces side bands spaced on either side of the carrier with difference frequencies equal to the frequency of modulation, so frequency modulation of the electron resonance absorption condition produces additional absorption lines as sidebands of the main line. Moreover these will be spaced on either side of the main absorption, with a difference frequency equal to that of the modulation frequency itself. If the frequency of modulation is low, in the audio region for instance, these side-bands will be extremely close to the main absorption and will, in fact, be buried right inside the line-width of the normal resonance line. If, however, the frequency of modulation is very high, the sidebands that are produced may be separated from the main absorption line, and they will then produce spurious spectra and limit the resolution of the system. The crucial factor is the ratio of the frequency of modulation to the normal line-width of the resonance absorption, expressed in frequency units. This relation between frequency and field, for the line-width, is exactly the same as the resonance condition itself, and it can therefore be seen that a modulation frequency of 1 Mc/s is equivalent to a magnetic field deviation of 0·357 gauss. A large number of electron resonance absorption lines have line-widths of the order of 1 gauss, or sometimes considerably less, and therefore even 1 Mc/s modulation frequency would produce noticeable broadening of the absorption lines and loss of resolution of any fine splittings.

As a compromise, therefore, between high sensitivity, which requires a modulation frequency as high as possible, and high resolution, which requires a frequency of modulation as low as possible, most modern electron resonance spectrometers employing high-frequency field modulation operate at a modulation frequency of 100 kc/s, which produces an additional line broadening of around 30 milligauss. If very fine lines are to be studied, as are sometimes obtained from free radicals in solution, even this modulation frequency is not acceptable. However, in a large number of cases, the line-widths observed are noticeably greater than 100 milligauss and in these situations the 100 kc/s modulation is quite acceptable.

If very narrow lines with line-widths of the order of 10 milligauss or less are to be expected, it is quite evident that high-frequency modulation of the magnetic field can no longer be tolerated and that some other means must be found to produce the high sensitivity associated with high frequencies of detection. Such a system is provided by the superheterodyne method of detection that is employed in all radar sets and in most normal radio-frequency receivers. In this system, a second klystron is employed to provide microwave power at a slightly different frequency from that used to produce the electron resonance absorption, and the two signals are then mixed to produce a beat frequency, or intermediate frequency, at the required value. The details of such superheterodyne spectrometer systems are discussed in the next chapter, and one such spectrometer is illustrated in Fig. 3.4. It may be noted here that the power from the local oscillator klystron, on the right of the figure, is taken and mixed with power from the signal klystron, which has passed through the resonant cavity system; and a difference frequency between these two is taken from the balanced crystal detectors to an intermediate frequency amplifier working at 30 mc/s. In this way, the crystals can be operating at their optimum detecting frequency, while the microwave absorption itself is not modulated at any high frequency at all. In fact, as is discussed in the next section, the frequency of modulation of the resonance absorption can now be at a low audio frequency, which will produce no noticeable line broadening whatever. In this way, both high resolution and high sensitivity can be achieved at the same time, but only at the expense of a much more complex spectrometer system.

It will be evident from the details given in this particular section, that a large number of the design problems of electron resonance spectrometers are concerned with methods to overcome the excess crystal noise, and it is the properties of the crystal which tend to dominate the design criteria throughout. It may be noticed, in this connection, that other detecting devices have been and are being tried, such as microwave bolometers and, more recently, photoconductive elements for work at the shorter wavelengths. At the moment, however, it would appear that the simple silicon or germanium semiconductor crystal is the most effective microwave detector, despite the various sources of excess noise associated with it and the somewhat complex methods by which this needs to be overcome.

References

1. Wilmshurst, T. H., *Electron Spin Resonance Spectrometers*, pp. 9 and 119 (Adam Hilger Ltd, London, 1968; Plenum Press, N.Y., 1968).
2. Wilmshurst, T. H., *op. cit.*, Chapter 6.
3. Wilmshurst, T. H., Gambling, W. A., and Ingram, D. J. E., *J. Electronic Control*, 1962, **13,** 339.
4. Ingram, D. J. E., *Free Radicals as studied by Electron Spin Resonance*, pp. 29–34 (Butterworths, London, 1958).
5. Poole, C. P., *Electron Spin Resonance: A Comprehensive Treatise on Experimental Techniques*, pp. 232–234 (Interscience Publishers Inc., New York, 1967).
   Wilmshurst, T. H., *op. cit.*
6. Poole, C. P., *op. cit.*
   Ingram, D. J. E., *Spectroscopy at Radio and Microwave Frequencies* (2nd edn, Butterworths, London, 1967).

Chapter 3

Spectrometer Systems

HIGH-SENSITIVITY SPECTROMETERS

3.1.1 *Elimination of low-frequency crystal noise*

The various features which have been discussed in the previous chapter can now be collected together and the actual design of a typical high-sensitivity electron resonance spectrometer can be considered. It has been seen that there are basically two forms of these; the first employs high-frequency field modulation, usually at 100 kc/s with a fairly straightforward microwave circuit, while the second employs superheterodyne principles, normally with a second microwave klystron and hence a much more complex microwave circuit. Although Fig. 2.6 suggests that the super-heterodyne system, operating at an intermediate frequency of 30 mc/s, should be noticeably less noisy than a spectrometer operating at a modulation frequency of 100 kc/s, the sensitivity of these two systems is not very significantly different in practice. The complex superheterodyne system is employed mainly because very high resolution is required to observe fine splittings in the spectra, not because high sensitivity is required. Brief descriptions of the two systems will now be given, but full details may be found in the texts listed at the end of the chapter.

3.1.2 *High-frequency field-modulation spectrometers*

A block diagram of a typical high-frequency field-modulation spectrometer is shown in Fig. 3.1. This employs a magic-T microwave element to produce the balanced bridge, as discussed in §2.3.2. Adjustment of the matching elements in the second arm of this bridge allows the power falling on the crystal in arm 4 to be varied independently of the power level in the absorption cavity itself. The high-frequency magnetic field modulation is provided from a separate 100 kc/s oscillator, which feeds a power amplifier, the output of which is connected to the modulating coils themselves. Since the skin depth of the copper or brass forming the cavity is very small at 100 kc/s, either the modulation coils will have to be put within the cavity and directly around the specimen, or else part of the cavity walls will have to be made extremely thin and the modulation coils mounted immediately behind them, so that the modulation field can penetrate the thin wall and still have a reasonable magnitude at the site of the specimen itself. Either of these methods of applying the 100 kc/s magnetic field modulation can be employed. In the case of a simple rectangular H_{102} resonator, the internal coil is often adopted, this taking the form of a few

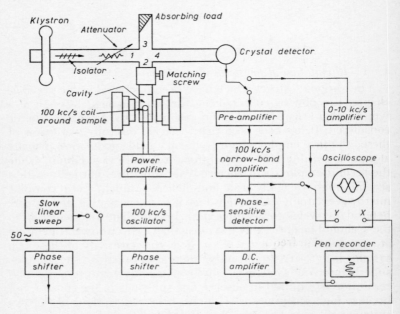

Fig. 3.1. Block diagram of high-frequency field modulation spectrometer
The modulation is provided by the 100 kc/s oscillator as shown, which also feeds a reference signal to the phase-sensitive detection equipment

loops of wire run through the cavity in a direction parallel to the microwave magnetic field lines and perpendicular to the electric field lines, as indicated in Fig. 3.2(*a*). Such loops can in fact be placed within the cavity, as shown, immediately around the tube containing the specimen; if they are carefully positioned with respect to the electric field pattern, no large reduction in cavity Q-factor will be produced. This single loop of wire represents a low impedance, and hence the output of the 100 kc/s amplifier must be designed to feed a high-current low-voltage signal to the modulation coil itself.

The alternative way in which the modulation can be applied, is shown in Fig. 3.2(*b*), which is illustrated for the case of a cylindrical resonant cavity. The central portion of the cavity wall is milled

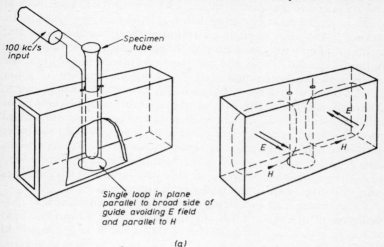

100 kc/s input

Specimen tube

Single loop in plane parallel to broad side of guide avoiding E field and parallel to H

(a)

100 kc/s input

Cylindrical cavity

Thin walled portion

Modulation coil in position

Basic design of coil

(b)

Fig. 3.2. Methods of producing 100 kc/s magnetic field modulation
(*a*) Internal coil within rectangular cavity
(*b*) External Helmholtz coils around thin-walled cavity

down until only a thin wall section is left over this central region; it must be sufficiently thin to allow the 100 kc/s field to penetrate, but sufficiently thick to maintain mechanical rigidity and also to act as a short circuit of continuous metal for the microwave frequencies. In fact the skin depth in copper for 100 kc/s is 0·2 mm and for microwave frequency of 9000 Mc/s is 7.10^{-4} mm; hence a wall thickness of one thousandth of an inch (0·02 mm) will meet both of these requirements. The modulation coils themselves can now be wound in the form of a curved Helmholtz pair, as indicated, and again supplied with power from an amplifier with a low-impedance output, as before. Once the coil has been placed in position, the cavity wall can be strengthened by casting in Araldite, or other similar resin, to give the whole assembly firm mechanical strength. Alternatively, the cavity can be initially fabricated of ceramic or other non-conducting material, when the coil can be wound around its outside and a thin layer of silver plating deposited on the inner surface to provide the resonant structure for the microwaves themselves. In practice, however, it is often difficult to produce a very smooth homogeneous surface in this way, and any irregularities will, of course, reduce the Q-factor of the cavity.

Either method of producing the 100 kc/s modulation will normally result in a fairly small amplitude of 100 kc/s magnetic field at the specimen, and magnitudes of the order of 20 gauss would normally be an upper limit. For this and other reasons associated with the long time constants of the detecting system, relatively small magnitudes of the 100 kc/s field modulation are normally employed and a first derivative rather than a straightforward absorption signal, is detected and displayed. The way in which this is done is illustrated in Fig. 3.3. Here the 100 kc/s magnetic field modulation is represented by the field sweep between the two vertical lines, and this is steadily moved across the profile of the actual absorption line by slowly changing the value of the main d.c. magnetic field, as indicated. As the line shape is swept through in this way, the field modulation samples the slope of the absorption line itself and produces an output signal which is proportional to this slope, as shown. It is the magnitude of this output signal which modulates the microwaves passed on to the detecting crystal, and the way in which this magnitude varies as the main d.c. field is swept through the resonance is indicated at the bottom of Fig. 3.3. The change in sign of this detected signal does, in fact, reflect a change in phase produced by the rising and falling slopes of the absorption line itself, and this change in phase is converted into a change of sign in the output signal by the use of the phase-sensitive detector circuits, as shown in Fig. 3.1.

The first derivative tracing illustrated at the bottom of Fig. 3.3 is, therefore, the way in which electron resonance spectra are normally presented. Such signals as these are traced out automatically by the pen recorder at the end of the spectrometer system, this being fed by the output of the phase-sensitive detector via a d.c. amplifier, as indicated. It is the bandwidth of this final amplifying and recording stage which is the ultimate limitation of the sensitivity of the spectrometer, as expressed by equation (2.31). High sensitivity can be obtained only by reducing the bandwidth of these circuits to as small a value as possible, and this, in turn, implies

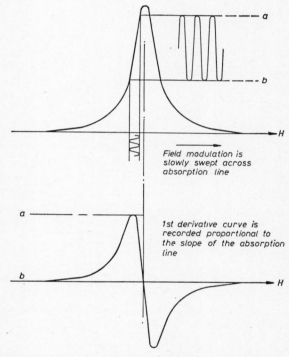

Fig. 3.3. First-derivative tracing produced by high-frequency field modulation

The way in which the small magnitude 100 kc/s modulation samples the profile of the absorption line is shown at the top and the first derivative tracing so produced is shown at the bottom

long time constants associated with them, and hence correspondingly long times for the actual sweep of the magnetic field through resonance. This, in turn, implies high stability of the microwave frequency and field homogeneity during the recording time. Ways in which this time constant can be effectively increased, and hence sensitivity also improved, are discussed further in §3.2.

3.1.3 *Superheterodyne spectrometer systems*

It has been pointed out already that the high-frequency field-modulation spectrometer will not give sufficient resolution if the widths of the electron resonance absorption lines are of 35 milligauss, or less. If such narrow lines are to be studied, it will be necessary to employ a superheterodyne method of detection, so as to avoid the modulation broadening which is produced by the high-

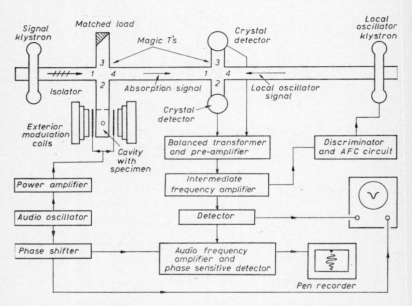

Fig. 3.4. Block diagram of superheterodyne E.S.R. spectrometer
The local oscillator klystron is shown on the right and the microwave signal from this mixes with the microwave power, carrying the absorption signal, in the balanced bridge containing the two crystals as shown

frequency modulation of the resonance condition itself. The block diagram of such a superheterodyne system is shown in Fig. 3.4. The microwave circuit on the left is basically the same as in the 100 kc/s spectrometer, and the third arm of the magic-T is balanced against the cavity, as before. The out-of-balance signal, carrying the electron resonance absorption, thus passes into arm 4, as in the previous spectrometer, but now, instead of being detected directly, it is mixed with microwave power from the local oscillator klystron. This is shown at the right of the figure, and operates at a frequency 30 Mc/s away from that of the signal klystron. The two microwave signals thus produce a beat or inter-mediate frequency at 30 Mc/s, which can be obtained from the two crystals mounted in opposing arms of the second magic-T system, as shown. The output from these two crystals is fed to a matching transformer, and the 30 Mc/s signal from this can then be fed to the main intermediate-frequency amplifier, as shown. The two detecting crystals are mounted in the second magic-T in order to eliminate any extra noise that might have been added by the local oscillator klystron. One of the features of the magic-T is that, whereas an input signal along arm 1 will produce signals of the same phase in the two balancing arms, the signal injected into arm 4 will produce signals in the balancing arms of equal magnitude but opposite phase. It is therefore possible, by suitably connecting the crystal outputs to the matching transformer, to cancel out any noise signals coming from the local oscillator, while at the same time adding the signals coming from the cavity absorption itself.

Since the detection of the microwaves is now taking place at this high frequency of 30 Mc/s, there is no need to use high-frequency modulation of the magnetic field, and audio modulation with negligible modulation broadening can therefore be employed to sweep through the actual resonance condition. This will produce an audio-modulated signal on the 30 Mc/s intermediate frequency, which can be detected in a phase-sensitive detector circuit, as before, and displayed by a pen recorder in the normal manner.

It is very important for the successful operation of such a super-heterodyne system that the local oscillator klystron has a frequency which is always exactly 30 Mc/s above, or below, that of the signal klystron. This can be ensured by incorporating automatic fre-quency control (A.F.C.) circuits, which take some of the output signal from the crystals and compare this beat frequency with those of standard tuned circuits, above and below the 30 Mc/s itself. A d.c. control voltage is then produced from these discriminating circuits and can be fed directly back to the reflector of the local oscillator klystron to change its frequency in the appropriate direction, if any drift should occur. Such automatic frequency control circuits are again normal design in lower-frequency super-heterodyne systems and can be taken over directly from them.

Most spectrometer systems, whether of the high-frequency field-modulation type, or of the superheterodyne type, also employ auto-matic frequency control circuits to lock the frequency of the signal klystron to the resonant frequency of the cavity containing the specimen. This kind of automatic frequency control can be achieved by the simple process of modulating the reflector of the signal klystron with a frequency of about 10 kc/s. This reflector

F

modulation then imposes a frequency modulation on the microwaves themselves as they leave the signal klystron, and such a
frequency modulation can be converted into an amplitude modulation by the Q curve of the cavity. The effect will be exactly the
same as that illustrated in Fig. 3.3 for the production of an
amplitude-modulated signal by the electron resonance absorption
line when the 100 kc/s magnetic field modulation passes across it.
In the case now being considered, however, the actual Q curve of
the cavity plotted against frequency acts as the effective absorption
line, and the frequency modulated microwaves are demodulated
by this, in the manner indicated in Fig. 3.5. If the cavity is exactly
in tune with the klystron, no asymmetric output signal will be produced, but if the klystron drifts away from the cavity, then an
amplitude modulation at 10 kc/s frequency will be produced on the
microwaves, and this can be detected as an additional signal, either
on the 100 kc/s signal of the field modulation spectrometer, or on
the intermediate frequency of the superheterodyne system. This
10 kc/s signal can then be filtered off from the main detecting and
amplifying run and fed back to a phase-sensitive detector, which is
also supplied with a reference signal from the original 10 kc/s
oscillator. A resultant d.c. voltage is then produced with a magnitude and sign reflecting the magnitude and direction of any drift
of the klystron from the centre frequency of the cavity. This d.c.
potential can, in turn, be fed back to the klystron reflector to offset
the drift and return the klystron frequency to the centre frequency
of the cavity. The way in which this is done is illustrated in block
diagrammatic form in Fig. 3.5.

The condition for maximum signal strength, and minimum distortion of the line shape, will be achieved by keeping the klystron
frequency locked to the centre frequency of the cavity, as described
in the previous paragraph. On the other hand, it is possible for the
cavity containing the specimen to drift and pull the klystron frequency with it. In this case, the maximum signal strength and
correct shape of the absorption line will still be obtained, but the
absolute value of microwave frequency at which the resonance
occurs may have altered from the value measured at the beginning
of the experiment. Correct determination of absolute g-values may
thus be in error, and if it is important that the g-values should be
obtained as accurately as possible, it will be better to lock the
frequency of the klystron to an external cavity, held in a constant
temperature enclosure, or to the harmonics of a crystal oscillator.
Very high Q reference cavities can be constructed to act as such
locking devices, and can be used to lock the klystron to their centre
frequency in exactly the same way as described in the previous

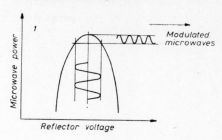

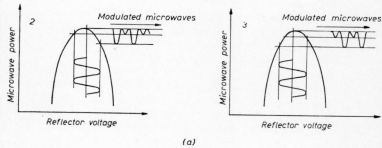

(a)

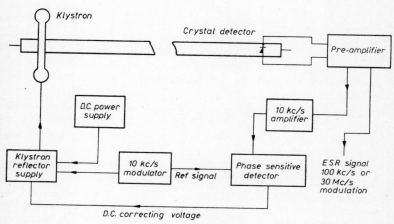

(b)

FIG. 3.5. Automatic frequency control

(a) Production of error signal. The variation of the microwave power with d.c. reflector voltage is shown as the inverted parabolic curve in each case. A constant modulation voltage is also shown applied to the reflector (1) when the klystron is in tune, (2) away from resonance on the higher voltage side, and (3) away from resonance on the lower voltage side

(b) Phase-detection of error signal to produce correcting voltage. The error signal is fed to a phase-sensitive detector, and the d.c. correction voltage from this is then fed back to the klystron reflector

paragraph for the cavity containing the specimen. It is thus possible to lock the signal klystron either to the cavity containing the sample under study, or to an external reference cavity, and the former method should be used if maximum sensitivity and faithful reproduction of line shape are required, while the latter arrangement is to be preferred if very accurate absolute measurements of *g*-values are required.

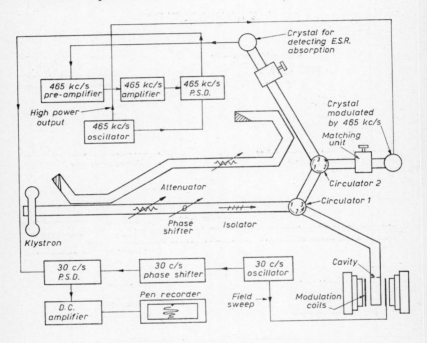

Fig. 3.6. Use of microwave circulators in E.S.R. spectrometers

The two circulators used in this microwave system, together with the crystal driven by the 465 kc/s oscillator, effectively produce a super-heterodyne spectrometer employing only one klystron

3.1.4   *The use of microwave circulators*

Both of the spectrometer systems outlined in the previous two sections have been illustrated with magic-T's incorporated in their microwave circuits. It was mentioned earlier, however, that microwave circulators are gradually replacing magic-T's, and one way in which this may be done is illustrated in Fig. 3.6. This figure also shows how superheterodyne methods of detection can be achieved without actually using a second klystron. Thus the power from the one klystron is taken via circulator 1 to the resonance cavity and then passed on to the circulator 2. A microwave crystal is mounted in one arm of circulator 2, as shown, and driven by an oscillator operating at 465 kc/s. The microwave power is thus modulated at this frequency and has sidebands impressed on it, which are displaced from the original frequency by ±465 kc/s. This frequency modulated signal now passes up arm 3 of circulator 2 to the detecting crystal mounted at its end. At the same time, a microwave bucking system takes some of the original microwave power, before any resonance absorption has modulated it, and feeds this back into the output arm of circulator 2, in the manner indicated. This microwave power coming through the bypass arm will now beat with one of the sideband signals of the modulated microwaves to produce an intermediate frequency of 465 kc/s in the crystal detector, which passes on through the narrow-band amplifiers, as shown.

The principle of superheterodyne detection is thus achieved by mixing the original microwave frequency with one that has been displaced 465 kc/s away from it; no automatic frequency control is needed to maintain this intermediate frequency at a steady value, since the oscillator modulating the crystal ensures this. It should also be noticed that no modulation broadening is produced by this method, since the resonance condition itself is not modulated at 465 kc/s, and the frequency of modulation applied to the magnetic field can be kept in the audio range, as in any normal superheterodyne arrangement. Two stages of phase sensitive detection take place in such a spectrometer system as this. The first detection occurs at the 465 kc/s frequency, where the intermediate frequency from the detecting crystal is compared with a reference signal from the oscillator driving the crystal, and the demodulated audio signal, representing the electron resonance absorption, then passes on to a second phase sensitive detector (P.S.D.), which is also fed with a signal from the audio oscillator. This system, therefore, combines the advantages of the superheterodyne system with the simplicity of the single klystron spectrometers, and is now being developed as a very successful commercial instrument. No further details of such systems will be given here, but readers requiring more information are referred to the texts listed at the end of the chapter.[1]

§3.2

AVERAGING AND INTEGRATING METHODS

Detection and display of very weak electron resonance signals has been revolutionized during the last few years by the introduction of averaging and integrating methods, and these will now be briefly discussed. The great power of these techniques is most dramatically shown in their ability to study rapidly changing spectra, and this application has particular significance in the study of active biochemical reactions. The ideas and methods are probably best

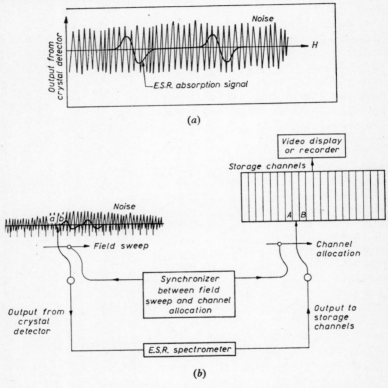

(a)

(b)

FIG. 3.7. Signal-to-noise improvement by integration techniques
(a) Typical signal buried underneath detector noise when only one spectral sweep has been recorded
(b) Division of magnetic field sweep into small increments, with their associated storage channels

approached first, however, by considering the way in which sensitivity for a static signal can be enhanced, and then following the argument through to transient phenomena.

The ideas can probably be best illustrated by considering a specific signal buried beneath the noise level, as indicated in Fig. 3.7(a). Thus, if the signal is traced out with a typical 10-second time constant, only random noise will appear on the pen-recorder trace, and it will be very difficult even to imagine the presence of any signal. The general principle of the averaging and integrating techniques is to take such a signal as this and repeat it many times, storing the output of each successive sweep and adding all successive signals to those already stored. This is represented schematically in Fig. 3.7(b), where the magnetic field axis is shown divided into a large number of small incremental values. As the d.c. field sweeps through these different small increments, so the output of the electron resonance spectrometer is fed successively into different storage channels. Thus, when the magnetic field has reached the particular incremental division between *a* and *b* on the diagram, the output from the spectrometer is being fed into a storage channel between *A* and *B* in the storage device. If there are, say, a hundred divisions of the field across the spectrum, there will be a hundred corresponding channels in the recording device to store the signal obtained from each incremental field position in turn.

If only one sweep through the resonance spectrum is stored in this way, the form of the stored signal will, of course, be exactly the same as that indicated in Fig. 3.7(a), and the actual signal will still be buried completely under the random noise fluctuations. If, however, the whole operation is repeated a second time, the magnitude of the spectrometer output fed into a given storage channel will again be determined by the level of the real signal strength, which will be exactly the same as for the first sweep plus the value of the random noise at the moment. Similarly for a third and fourth sweep, each channel will receive an input composed each time of the actual absorption signal strength, corresponding to its particular field increment, plus the random noise that exists at the particular moment. The stored magnitude of the actual signal, corresponding to the resonance absorption itself, will add linearly with the number of sweeps, n, that are fed into the storage device, but the value of the random noise on any sweep will be quite unrelated to its value on any other, so that the noise components will sum in a way corresponding to the random addition of signals, which can be shown to be proportional to $\sqrt{n}$, where n is the number of sweeps.

It, therefore, follows that if n sweeps have been fed into the storage device, the stored signal due to the electron resonance absorption, will be directly proportional to n itself, whereas the noise stored will be proportional to $\sqrt{n}$. Thus the signal-to-noise ratio will grow continuously as the number of sweeps increases and will, in fact, grow as $\sqrt{n}$. In this way, the actual signal strength itself can be integrated out of the noise background; and if the output of the storage device is displayed on an oscilloscope, or pen recorder, the signal will become discernible above the background noise after a requisite number of sweeps have been fed into the system.

There are two essentials for any such integrating system. The first is a suitable storage device and the second a synchronizing system. Synchronization is necessary so that the field sweep can always return to start at the same initial point, and the successive storage channels must open at the right moment. In the early study

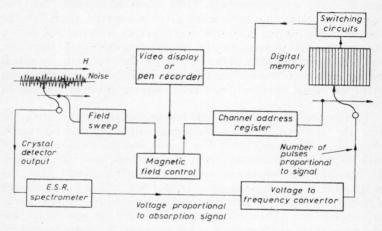

FIG. 3.8. Block diagram of integrating system employing digital memory as storage
In this the E.S.R. signal strength is converted into a series of pulses by the voltage-to-frequency converter and the signal strength is thus stored as the number of pulses in the manner indicated

of such possibilities, many storage devices were tried, such as magnetic tapes and photo-cathode tubes, but the methods received a great impetus when workers such as Klein and Barton[2] showed that a digital computer could be used very effectively as such a device. The computer can be regarded as a form of digital memory composed of numerous different channels, with each channel capable of storing a number which can be changed from zero to 10^4. Since this particular kind of memory stores numbers, rather than summing input voltages, some device must be placed between the output of the electron resonance spectrometer and the input to the storage channels themselves. Thus the output of the spectrometer is basically a voltage which measures the strength of the absorption signal plus noise, and this is converted into an actual number, proportional to the magnitude of the voltage, in a voltage-to-frequency converter, as indicated in Fig. 3.8.

Synchronizing circuits are also included, so that, as the electron resonance spectrum itself is swept through by the varying magnetic field, the output from the spectrometer is successively fed into the different storage channels. The numbers fed to any channel are then directly proportional to the strength of the output signal of the spectrometer at that particular instant. The whole process is then repeated as many times as required, and the signal-to-noise ratio is steadily improved within the memory system, in the way outlined above. The effectiveness of such integrating techniques can be seen from Fig. 3.9, where the first example (*a*) is for a spectrum originally obtained by Klein and Barton[2] from a weak solution of manganese ions in water. As mentioned previously, this spectrum is often obtained in biochemical studies. A single sweep gives no sign at all of the absorption, whereas, after 5000 integrated sweeps, the signal is extremely clear and the noise has almost completely vanished. In the second example (*b*), a free-radical signal is integrated out of the noise background in a similar way and the successive improvement of the signal-to-noise ratio can be clearly seen; even a quantitative dependence on $\sqrt{n}$ is evident in the results.

Computer methods of averaging and integrating such static signals are so powerful that they are now almost accepted as a necessary accessory to any high-sensitivity electron resonance spectrometer. Such a device employing digital memory circuits for storage purposes is technically called a computer of average transients, but is known as a CAT for short.

§3.3

THE INTEGRATION OF RAPIDLY CHANGING SPECTRA

The use of integrating techniques, as described in the last section, for the enhancement of static signals can be very impressive, but it does not really introduce any new principle into the detecting methods already discussed. Thus, higher sensitivity has been achieved effectively by introducing a much longer time constant, and hence a very much narrower recording bandwidth, into the detecting system. This time constant has been obtained by summing a large number of smaller time constants, which are fed successively into the recording device; these smaller times are the times taken to sweep through each individual spectrum. The general conditions of stability for both the microwave frequency and the field controls still apply however, and, in principle, the

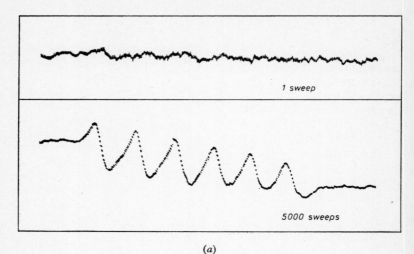

1 sweep

5000 sweeps

(a)

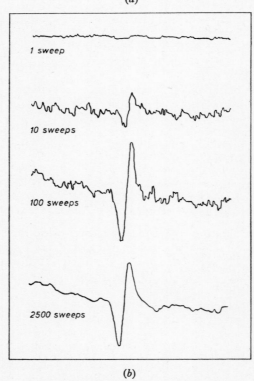

1 sweep

10 sweeps

100 sweeps

2500 sweeps

(b)

FIG. 3.9. Typical integrated E.S.R. signals
(a) Spectrum obtained from a weak solution of manganese ions in water
(b) Spectrum obtained from a small quantity of a typical organic free radical

same sensitivity for a static signal can be achieved either by taking a hundred successive 10-second sweeps through a spectrum and feeding these successively into a CAT or by just sweeping through the spectrum once, but taking 1000 seconds to do it and employing extremely narrow bandwidths with long time constants in all the detector circuits. In practice, the former method is the easier to achieve, and this is why CATs are so useful in enhancing the sensitivity of spectrometers which are used to study static signals.

The investigation of rapidly changing electron resonance spectra produces quite a new concept, however. With such spectra, the integrating method can actually achieve results which would be quite impossible if the method were not available. If a spectrometer system containing no integrating device is considered first, it will be seen that there is now a fundamental conflict between the requirement for high sensitivity and the requirement for a rapid recording of changing signal strength. Thus we have already seen that high sensitivity will only be achieved if very narrow bandwidths are used in the detector circuits, but the amplification and display of rapidly changing signals imply exactly the opposite of this, i.e. wide bandwidths for the appropriate amplifying and recording circuits. It would, therefore, appear at first sight that it will be quite impossible to record rapidly changing electron resonance spectra unless these are of very high intensity, and hence do not require the narrow bandwidth of a high-sensitivity system. This statement is just one example of a fundamental theorem of information theory which, in this context, can be phrased in the form 'the sensitivity of any recording system will decrease as the time taken in recording the information is decreased'.

The situation can probably be best illustrated by taking the example illustrated in Fig. 3.10(a), which represents the growth and decay of the centre of an electron resonance absorption line as a biochemical reaction proceeds. Thus the signal is seen to grow initially, as the interaction takes place, and then to decay as a secondary action takes over. If this changing signal were to be followed by the electron resonance spectrometer directly, the magnetic field and microwave frequency could be held on the centre of the absorption line, and, if a large absorption were being produced, its change could be followed directly as a change of the spectrometer output with time. Thus the actual curve indicated in Fig. 3.10(a) could be traced out on an oscilloscope screen, or some other high-frequency recording device, where the vertical axis represents the instantaneous strength of the resonance absorption and the horizontal axis the time. Such a direct plot of the changing signal could be obtained, however, only if wide-band amplifying

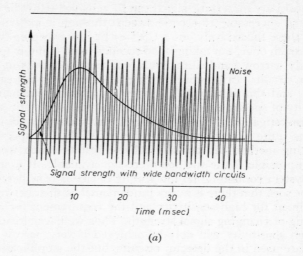

(a)

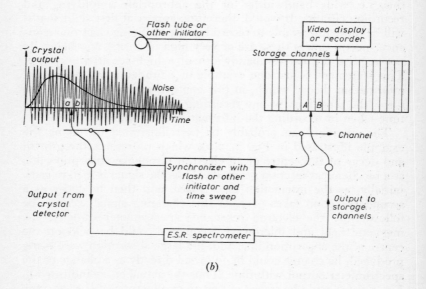

(b)

FIG. 3.10. Integration of rapidly changing spectra
(a) Typical time variation of E.S.R. spectrum from a biochemical reaction
(b) Block diagram indicating division of time scale into successive increments
and corresponding storage arrangements

and recording circuits were used, and the large bandwidths would be the source of a large amount of additional noise. Thus, if the signal were only from a small amount of enzyme, or other reacting specimen, the signal would be buried under the random noise of the amplifying system and no recognizable recording would be obtained.

If a particular transient action can be repeated, however, exactly the same argument can be applied to its recording as was applied to the recording of a static signal, but with the magnetic field axis of the recording system replaced by the time axis. Thus, if a photochemical reaction is taken for simplicity, and the variation of Fig. 3.10(a) represents the growth and decay of a particular photochemically induced species, the time axis can be divided into segments just as the field axis was before, and each part of the time variation can be successively fed to storage channels as in the previous case. The synchronization now takes place between the time scale and the sequential opening of the storage channels, and it is initiated in each instance by the photochemical flash, defining the zero time for each successive sweep. In this way, the signal strength obtained at any instant of time after the initial flash can be gradually built up within a given storage channel. The whole time variation can thus be integrated out of the random noise background, since it will again increase linearly with the number of sweeps, whereas the noise will increase only with the square root. It therefore follows that the principles used for enhancing a static signal can be used to enhance the time-variation pattern of a rapidly changing signal, as indicated in Fig. 3.10(b).

It is, of course, possible to study and to obtain permanent records of both the variation in time of a rapidly changing signal and also its variation with field. Fig. 3.11 is a three-dimensional perspective representation of all the information that can be obtained about a rapidly changing electron resonance spectrum, which, for purposes of illustration, is shown with triplet splitting; other absorption patterns could be shown similarly.

At any given instant of time, the spectrum might have a profile in its magnetic-field variation, as represented by PQ on the diagram, but it would be difficult to trace this particular hyperfine pattern in detail, since it would change during the process of recording. Instead, the techniques described in the last paragraphs can be employed to trace the time profile of each magnetic-field value in turn. Thus, the slice AB represents the time variation of the maximum of the electron resonance pattern, and this can be obtained from one set of integrated sweeps, as described above. Once this information has been obtained and recorded, it is then possible to

change the setting of the d.c. magnetic field and obtain a second
time profile at this new field position, as indicated by the slice CD.
It will be evident from this that the whole three-dimensional
pattern of information can be built up from several successive
integrated sweep operations, and the complete time variation of
even a complex spectrum obtained in this way.

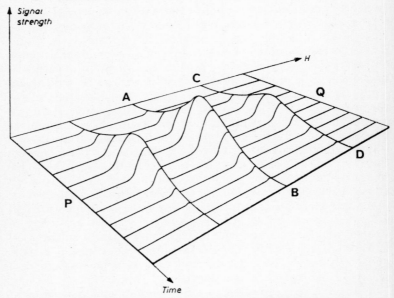

FIG. 3.11. Total variation of spectrum in time and field strength as
represented on a three-dimensional plot

Some types of computer integrating circuits are now being
developed which enable simultaneous integration of both the
time and field variation, so that the spectrum observed at any
nstant can be immediately displayed on the computer output.

§3.4

THE STUDY OF TRANSIENT REACTIONS

3.4.1   *Methods of studying non-repetitive transient reactions*

In the previous two sections, the methods whereby the sensitivity of electron resonance spectrometers can be enhanced by
averaging and integrating techniques have been discussed in detail
for both static signals and rapidly changing signals which can be

repeated a large number of times. The situation may often arise, however, especially in biochemical or biological studies, where it is impossible to repeat any given reaction exactly, so that the integrating methods described above will no longer be applicable. Other methods must therefore be devised whereby non-repetitive transient reactions can also be studied and their kinetics analysed.

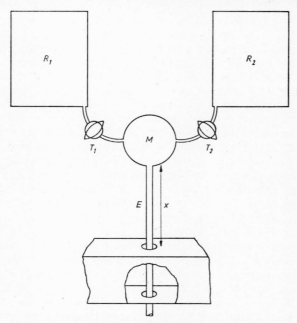

FIG. 3.12. Block diagram illustrating general principles
of continuous-flow systems

The reacting solutions are mixed together in the chamber M and the time between the moment of mixing and the moment of E.S.R. resonance absorption can be varied by altering the distance x

A method can be devised fairly simply for any system in which the reaction in question can be made to take place in a moving stream of gas or liquid in such a way that the dynamic concentration of the transient species changes down the length of the flow tube. The concentration at different points of the tube will then reflect an actual change of the kinetics with time. This principle can probably be best illustrated by the general example illustrated in Fig. 3.12.

In this example it is assumed that the reaction of interest, which may, for instance, be the interaction of an enzyme with its substrate, does not take place until two separate solutions are brought together in a mixing chamber. Thus, the two initial components can be stored separately in the reservoirs denoted by R_1 and R_2 and only brought into contact when the taps, T_1 and T_2, leading to the mixing chamber, M, are opened. The two solutions are then thoroughly mixed in this mixing chamber and pass along the exit tube, E, which itself passes through the rectangular waveguide forming the resonant cavity of the spectrometer system. If the steady rate of flow of these two reactants into the mixing chamber is maintained, there will also be a steady flow of the resultant mixture through the exit tube, E. It, therefore, follows that each point along this tube, as measured by the distance x, will also correspond to a specific time of elapse from the beginning of the reaction when the two solutions first met in the mixing chamber. Different dynamic concentrations of the transient intermediates will, therefore, be set up along the exit tube, E, and a measurement of these differing concentrations with varying values of x, can be translated into a kinetic measurement of the varying concentration with time. It is thus possible, in principle, to study this varying concentration with time, by simply altering the distance between the mixing chamber and the electron resonance cavity itself, and noting the different absorption signal strengths that are obtained for the successive values of this distance of separation.

This general principle has, in fact, been developed along two main lines. The first of these lines follows the general description outlined above, and a number of different cavity resonators have been designed so that a reacting solution can be passed through them in the manner indicated; and further details of these are given below. The second line of development has been a slight modification of this principle, in that, instead of passing directly through the cavity, the reacting solutions are suddenly deep frozen at a predetermined distance along the tube, and thus at a predetermined time after mixing. This deep freezing completely quenches the reaction, and the frozen solutions can then be studied at leisure in the electron resonance spectrometer afterwards. This technique, known as the sudden freezing technique, is also discussed below.

3.4.2 *Continuous flow systems*

An example of how the general principle of the previous paragraph can be applied is shown in the flow apparatus illustrated in Fig. 3.13. This follows the general type of design initiated by Piette,[3] Bray[4] and others. The two initial solutions are stored in the syringes at the top of the apparatus; the two syringes differ in cross section and capacity, since the reactants are not to be mixed in equal amounts. Driving plungers down the two cylinders at an equal rate then produces the correct proportions for mixing in the mixing chamber itself. This mixing chamber, seen below the two reservoir syringes, consists basically of a group of fine jets, which set up a turbulent system of vortices in the chamber and ensure rapid mixing of the two solutions as soon as they enter it. The quartz exit tube leads down from the chamber and the

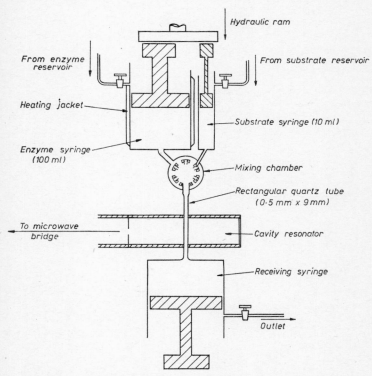

FIG. 3.13. Schematic diagram of practical continuous-flow system

electron resonance cavity is placed around the tube. The distance from the centre of this cavity to the mixing chamber can be varied, and hence the time variation of the kinetics can also be studied. It is, of course, also possible, as an alternative, to leave the distance between the mixing chamber and the cavity fixed and alter the rate at which the plungers are driven down the reservoirs, and hence also the rate at which the reacting solution travels through the quartz tube. Since dynamic concentrations of the

G

transient compounds are being produced in this way, the actual
concentration of the intermediate paramagnetic species will remain
constant at any given position along the tube, as long as the driving
plungers of the syringes are being driven down at the same rate.
It is therefore possible to use a fairly long time constant for the
detection and display of this signal, and in this way high sensi-

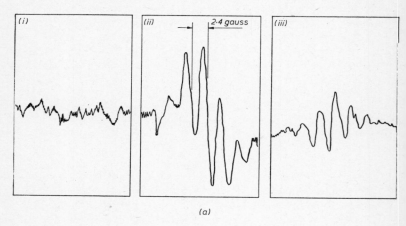

(a)

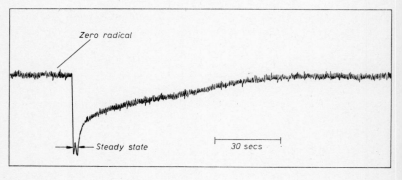

(b)

Fig. 3.14. Spectra from continuous-flow systems

(a) Formation of semiquinone radical in the presence of peroxidase as
a catalyst. In each case 10^{-2} M hydroquinone and 10^{-2} M H_2O_2 were
flowing through the cavity at a rate of 8 ml per second. In (i) no peroxidase was added, in (ii) 4×10^{-8} M peroxidase was added, and in (iii)
this was reduced to 1×10^{-8} M

(b) Decay curve of ascorbic acid free-radical. The spectrometer was
set on the maximum signal and the flow rate then stopped so that the
decay of radical concentration with time could be directly measured
(From Yamazaki, Mason, and Piette[3])

tivity can be achieved. The rapid time-variation of the signal has now been converted into a change with distance along the exit tube, and variation with this parameter can be studied at leisure.

Examples of spectra that have been obtained from such continuous-flow systems are shown in Fig. 3.14, and it is evident from these that fairly low concentrations of paramagnetic species can be detected in this way, even though they correspond to rapidly changing transient intermediates.

The great disadvantage of this particular system is that it tends to consume large quantities of the reacting solutions and can consequently be applied readily to biochemical or biological systems only when large amounts of the material under study can be obtained. If, however, only relatively small quantities of the materials are available, as may well be the case in the large number of biochemical studies, the technique needs to be modified so that maximum use is made of all the material passing through the system, as in the method described below.

3.4.3 *Sudden freezing techniques*

The sudden freezing technique for studying the kinetics of transient intermediates is based on the same general principle as the more direct continuous flow system, but instead of passing the reacting solution through the cavity resonator itself, it is suddenly injected into a region of very low temperature. There it is deep frozen with its transient concentration trapped and permanently equal to that existing at the moment of deep freezing. Such a system is outlined in Fig. 3.15. The first part of the equipment is the same as in the simpler continuous-flow system described above, and the reacting solutions are again driven by syringes into a mixing chamber, from which the interacting solution passes down the reaction tube. It then passes through the fine jet and is immediately deep frozen. The essential requirement in this technique is that the quenching process should be rapid in comparison with the time of the reaction being studied. Such rapid quenching and deep freezing can be achieved by injecting the reacting solution into a liquid in which neither of the reactants themselves is miscible and which is itself held at liquid nitrogen temperatures. It is in fact better to use a buffer liquid for this purpose, rather than liquid nitrogen itself, since production of gas bubbles throughout liquid nitrogen tends to retard the heat transfer from the reacting solution to the cold nitrogen liquid. The effectiveness of this rapid quenching has been studied in detail by Bray,[4] who showed that it will take only about 10 milliseconds for an aqueous solution, originally at room temperature, to become deep frozen after injection through a jet, with a nozzle diameter of 0·2 mm, into hexane held at −80°C. It, therefore, follows that this sudden freezing technique can be quite effectively used to study the kinetics of any enzyme reactions, or similar processes, where the reaction times for the transient intermediates are likely to be 50 milliseconds or longer.

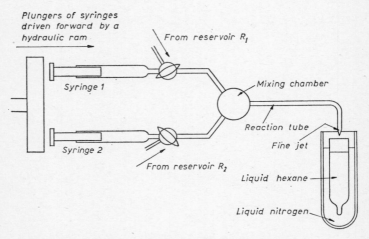

FIG. 3.15. Block diagram of sudden-freezing technique
The reacting solutions are mixed as for the continuous-flow method, but are then instantaneously deep-frozen after passage through the jet and are studied at leisure later

An example of the kinetics of a biochemical system that has been studied by this technique is given in Fig. 3.16, which shows a series of results obtained for the reaction of xanthine oxidase.[5] In this particular reaction, three different valency changes are produced in the transition group ions associated with the enzyme, while a high concentration of free radicals comes into transient existence at the same time. The kinetics of the valency changes, and their comparison with the free-radical concentration, are plotted together in Fig. 3.16. This set of results is a good example of the great power of electron resonance to correlate free radical activity with simultaneous valency changes of complexes within the enzyme. Detailed discussion of the implication of these results will be reserved until Chapter 5, but the particular results are quoted here as a good example of the sudden freezing technique in action.

The best results will be obtained with this technique if one or two precise experimental requirements are fulfilled. Thus, the flow of the liquids into the mixing chamber should be accelerated very rapidly at the start of the experiment to a predetermined value, and then held at this constant value until sufficient of the reacting material has passed through the jet into the cold liquid. The flow should then be immediately stopped, so that all the material deep frozen in the cold container corresponds to the same passage of time down the exit tube. Automatic mechanical systems employing hydraulic rams or similar devices should therefore be used to

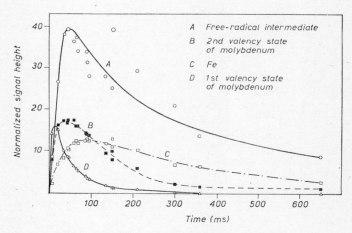

FIG. 3.16. Kinetic studies of xanthine oxidase
obtained by the rapid-freezing technique

The changing concentration of the different metal ions together with that of the specific free radical transient can be clearly seen (After Bray *et al.*[5])

ensure accuracy and reproducibility of syringe displacement, since any error in this will directly affect the accuracy of the kinetic measurements themselves. The cold hexane, or other liquid, into which the acting solutions are injected, can itself be held in a tube with a narrow tail some 3 mm in diameter, so that when the deep-frozen material has collected in this container, it can be packed down into the narrow tail, and this can then be inserted directly into the electron resonance spectrometer. It is a technique which tends to be employed in all cases where the supply of the material under study is limited, and the operation at low temperature not only enables the quenched material to be studied at leisure, but also gives some extra sensitivity from the low-temperature measurements themselves.

§3.5

ZERO-FIELD SPECTROMETERS

It has been tacitly assumed in all the previous discussions that the magnitude of the microwave quantum used to induce the electron transitions, was sufficiently great to supply the energy required to jump from one level to the next. If there are an odd number of unpaired electrons associated with the atom or molecule under study, this indeed will always be true, since there is a general theorem due to Kramers which states that no internal crystalline electric field can completely remove the degeneracy of a system containing an odd number of unpaired electrons. Thus each level will at least be twofold degenerate, and this final degeneracy can be removed only by the application of the externally applied magnetic field. A good example of this has already been illustrated in Fig. 1.7, where it was seen that the internal fields did, in fact, remove some of the degeneracy of the six spin levels of Mn^{++}, but because there were an odd number of unpaired electrons (5 in this case), each resultant level was still left doubly degenerate in spin. This final degeneracy can be removed only by the application of the magnetic field.

If, however, there are an even number of unpaired electrons in the system, the internal field can remove all the degeneracy from the electronic levels, and thus produce a splitting between the energy levels, even in zero applied magnetic field. As a simple illustration of this, the zero-field splittings that are obtained for an atom containing two unpaired electrons, placed in a non-axially symmetric electric field of a crystal lattice, are shown in Fig. 3.17. This may be compared with the somewhat simpler Fig. 1.5(a), where the zero-field splitting was produced by an axially symmetric electric field. This resolved the degeneracy between the $M_s = \pm 1$ levels and the $M_s = 0$ level, but did not separate the ± 1 levels themselves. It will be seen, however, that in the more general case of Fig. 3.17, where all three levels are separated, electron resonance absorption will be possible only if the magnitude of the microwave quantum is greater than the initial zero-field splittings, or if the magnetic field can be increased to such a value that two of the levels are brought within the magnitude of the microwave quantum. In quite a large number of cases, the zero-field splittings are quite small, and therefore no question arises whether the microwave quanta are large enough. However, in other cases, the zero-field splittings may be noticeably larger than the value of the microwave quantum, and then there will be no guarantee that any electron resonance absorption will be observable with the magnetic field strengths available. It is of interest that some of the particularly important biochemical compounds, such as haemoglobin, are among those which have abnormally large zero-field splittings, and the particular reasons for this are

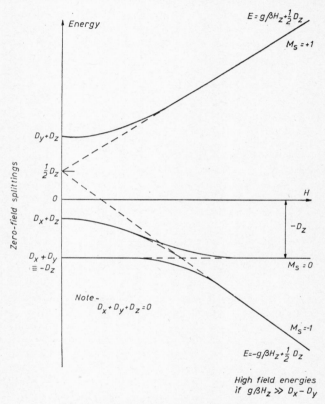

FIG. 3.17. Electronic zero-field splittings
in non-axially symmetric electric fields

The degeneracy of the $M_s = \pm 1$ levels is now resolved by the lower symmetry field and the three energy levels have different values even in the absence of the applied magnetic field

undoubtedly related to the structure and function of the molecule itself.

The accurate determination of these zero-field splittings is very important if a detailed picture of the state of the atoms and molecule concerned is to be obtained, and hence some considerable effort has recently been expended in developing electron resonance spectrometers which can measure these zero-field splittings directly. If the zero-field splitting itself is still too large to be spanned by the microwave quantum, it may be possible to bring the transitions within reach of the microwave quanta by the application of very large magnetic fields. In this connection, considerable interest has also been growing of recent years in the development of E.S.R. spectrometers for very short wavelengths, operating at very high field values, such as those that can now be obtained from superconducting magnet systems. A brief general description will therefore first be given of zero-field spectrometers, in which no external magnetic field is applied, and which may be used to measure the actual zero-field splitting directly. This is then followed by a short description of some of the more recent millimeter-wavelength spectrometers, operating with superconducting magnets.

The general features of a direct zero-field spectrometer are illustrated in Fig. 3.18, and there are two basic changes from all spectrometer systems that have so far been discussed. In the first place, it will be necessary to replace the klystron with a microwave oscillator which has a much wider tuning range. The microwave frequency will now be the only variable parameter in the experiment, and hence it must be variable over as wide a range as possible. Two types of device can be used for such a microwave source, either a backward-wave oscillator, which uses the principle of velocity modulation in a helix structure and contains no high-Q resonant cavity systems, or other frequency-determining elements. A backward-wave oscillator is included in the spectrometer of Fig. 3.18, and the frequency of the microwaves it produces is controlled by the voltages from the stabilized power supplies. An alternative possibility, now being developed, is to employ the harmonics from solid-state oscillators, operating at somewhat lower frequencies and tunable over a sufficient range to give complete frequency coverage from one harmonic to the next.

The other basic difference from previous spectrometers is that it will now be impossible to include a cavity resonator or other frequency sensitive device in the microwave circuit. The sample is therefore placed directly in the main waveguide run itself, and is mounted along one of the narrow sides of the waveguide, as shown in the diagram, so that it is still located in the region swept through by the maximum microwave magnetic field, but avoids the electric field component. Although no d.c. magnetic fields are being applied in such a zero-field measurement, a small a.c. modulation can still be used to sample the resonance line and provide a system of phase-sensitive detection for the final display of the signal. The absence of the concentrated microwave power in a cavity, and also of precise balancing systems, reduces the sensitivity of such a spectrometer to a value somewhat below that of the normal systems previously discussed, but if fairly concentrated specimens are being studied in the form of single crystals this does not usually

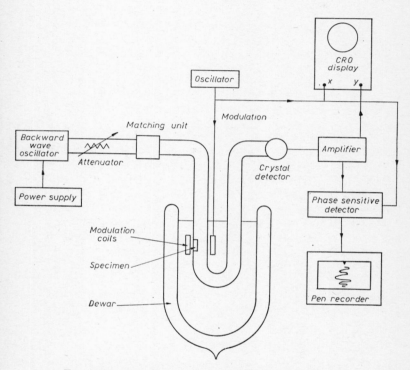

FIG. 3.18. Block diagram of zero-field E.S.R. spectrometer
A variable frequency backward-wave oscillator is now used as a micro-wave source and all resonant systems are removed from the microwave circuit

matter. Details of specific spectrometers employing these general principles may be found in the original papers of Mock,[6] and Bogle, Symmons, Burgess and Sierins.[7] If higher sensitivity is required in such zero-field studies, as may often be the case in biochemical work, it is probably better to work with a normal form of spectrometer system, but operating in the millimetre wavelength range, and rely on a large d.c. magnetic field variation to cover the energy differences required.

§3.6

MILLIMETRE WAVE SPECTROMETERS

3.6.1   *Magnetic field requirements*

Electron resonance spectrometers which operate in the milli-metre wavelength region have the same basic design features as those working at lower frequencies, but there are one or two parti-cular points that need careful attention at these shorter wavelengths. If the spectrometer is to be used to study ordinary electron reson-ance spectra, centred on g-values of 2·0, as well as for any zero-field studies, it must be equipped with a magnet able to produce the required field value. For a spectrometer working at 4 mm wave-length, the resonance magnetic field for $g = 2$ will be about 26 000 gauss; for the 2-mm-wavelength spectrometer, the re-quired field value will be about 52 000 gauss. Until quite recently, it seemed impossible to produce continuous fields of this order of magnitude in any magnet system of reasonable size, but the advent of superconducting magnets has made the matter fairly simple and straightforward. These magnets consist basically of a solenoid wound of superconducting wire and continuously im-mersed in liquid helium. The cavity containing the specimen is placed down the central axis of the solenoid. Either the cavity is immersed directly in the liquid helium used to cool the solenoid or, if studies are to be made at higher temperatures or at room tem-perature, the dewar containing the solenoid is fitted with a re-entrant column so that the sample can be maintained at room temperature or any other suitable value.

3.6.2   *4-mm spectrometer system*

A block diagram of a 4-mm-wavelength spectrometer, employing such a superconducting magnet system, is given in Fig. 3.19. The microwave power is supplied from the klystron shown on the left of the figure, and most of the power it emits is fed to the central magic-T, as usual; the cavity resonator is placed at the end of the third arm. One particular feature may be noted here: the use of the oversize waveguide to feed the microwave power from the magic-T itself to the cavity, which is located in the centre of the solenoid and immersed in liquid helium, or in the centre of a re-entrant dewar if this is employed. There is, therefore, a fair length of waveguide run from the magic-T to the cavity resonator, and, if the normal size of waveguide for 4-mm-wavelength radiation is employed, considerable attenuation of the microwaves will result. It is therefore much better to insert two tapers which allow a larger, oversize, waveguide to be used for this main waveguide run, since, although the tapers introduce a small amount of

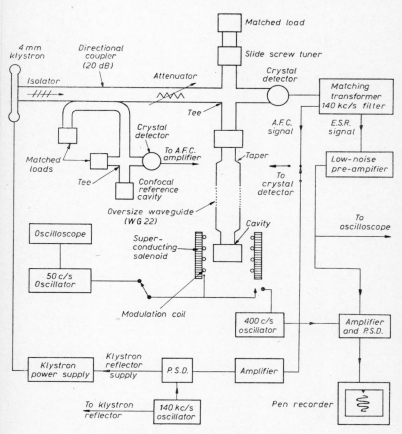

FIG. 3.19. Block diagram of 4-mm spectrometer

The superconducting solenoid and microwave cavity are actually immersed in a multi-wall dewar holding liquid helium

attenuation themselves, it is only a fraction of that introduced by a complete length of standard 4-mm waveguide. These points are of some importance in the high-frequency spectrometer system, since the microwave sources have limited power output in this wavelength region.

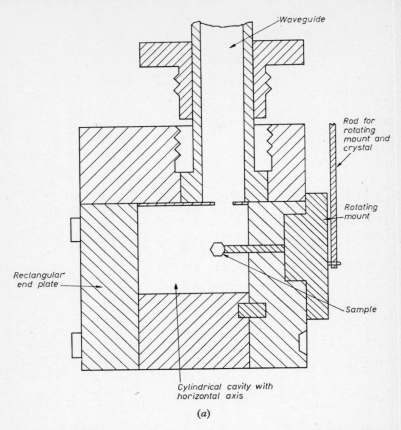

Waveguide

Rod for rotating mount and crystal

Rotating mount

Rectangular end plate

Sample

Cylindrical cavity with horizontal axis

(a)

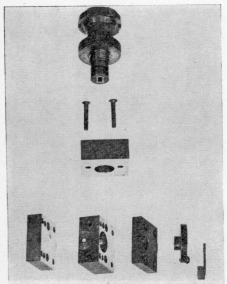

(b)

(c)

The out-of-balance signal produced by the electron resonance absorption is detected as usual by a crystal mounted in the fourth arm in the magic-T bridge, and is then passed to the various amplifier and detecting circuits.

In this particular spectrometer, the klystron frequency is locked to an external reference cavity rather than to the cavity containing the specimen, since *g*-values are to be measured as accurately as possible. This reference cavity can take the form of a Fabry-Perot confocal resonance system, in which two spherical mirrors form the two end surfaces of the cavity, which can be regarded effectively as an optical resonance system. Such confocal cavities can be fabricated with extremely high Q-values and thus form ideal cavities for frequency-locking purposes. The 140 kc/s modulation frequency, which is applied to the klystron reflector, is thus demodulated by this cavity; and any drift of the klystron from the cavity frequency will produce a d.c. correcting voltage, after mixing in the 140 kc/s phase-sensitive detector, as shown.

This particular spectrometer system was designed specifically for the study of haemoglobin crystals, which have line-widths of the order of 100 gauss or more. It was therefore not sensible to employ 100 kc/s magnetic-field modulation, since the amplitude of the modulation would be only a small fraction of the line-width, and so a magnetic-field modulation of 400 c/s was employed instead. This can be provided by a modulation coil mounted directly inside the coil of the superconducting magnet, as shown. The oscillator which feeds this coil also supplies a reference signal to the output phase-sensitive detector, which in turn passes the final detected signal to a pen-recorder.

Two other features of this spectrometer system deserve some mention. One is the design of the specimen cavity itself; the other is the alternative microwave detector that can be employed. A slightly novel feature about the specimen cavity is that it has to be mounted with its axis at right angles to the normal direction. Thus the d.c. field provided by the superconducting magnet will be along the axis of the solenoid and thus normally in the vertical direction. For maximum absorption, the direction of the microwave magnetic field must always be at right angles to the d.c. field, and hence a cylindrical H_{011} cavity will need to be mounted so that

---

Fig. 3.20. (*facing*) Microwave cavity for use with superconducting magnet

    (*a*) Cross-sectional diagram

    (*b*) Photograph of cavity components disassembled and laid out in an expanded form

    (*c*) Photograph of assembled cavity compared with threepenny piece for size

its cylindrical axis is horizontal. Such a cavity design is shown in Fig. 3.20(a), and it can be seen that the microwave power is now coupled in along one vertical side of the cylindrical cavity and that the actual coupling to the cavity can be varied by rotating the long face of the waveguide about a vertical axis at right angles to the axis of the cavity. This feature enables optimum coupling to be obtained in a very precise manner. The cavity is sealed by a rectangular end plate, as shown. A photograph of the system in an expanded form and compared with a threepenny piece for size is shown in Fig. 3.20(b) and (c).

The other particular point that may be noted in connection with the spectrometer is the use of an indium antimonide detector in place of the normal crystal detector. These indium antimonide detectors have been developed for use in the far infra-red region, and at 4 mm the microwave wavelengths are becoming sufficiently short for them to be used effectively in such an electron resonance spectrometer. Indium antimonide detectors depend on a photo-conductive effect and have to be cooled to liquid helium temperatures and operated in a high magnetic field. The magnetic field strength and the d.c. bias current, flowing through the indium antimonide crystal itself, are varied independently to obtain maximum sensitivity for the microwave detection, and magnetic fields of the order of 10 kilogauss and bias currents of about 70 μA are normally required. It will be appreciated that this kind of detecting device requires a considerable amount of ancillary apparatus and therefore is only to be justified if the maximum sensitivity is required. It does, however, increase the observed signal-to-noise ratio by more than a factor of 2, as is illustrated in Fig. 3.21, which compares (i) the electron resonance signal observed in a 4-mm-wavelength spectrometer with an indium antimonide detector with (ii) exactly the same signal taken under identical conditions, but with a silicon crystal as detector.

The overall sensitivity of this type of spectrometer system can be measured by using standard samples, such as silicon doped with phosphorus containing a known number of free spins. It has been found that the minimum number of spins which can be detected is of the order of 3×10^{10} per mW available power per gauss line width. The signal actually obtained from a sample containing 4×10^{12} spins is shown in Fig. 3.21(b). The cavity Q-factor was about 5000 for this measurement, the time constant for the recording 1 second, the microwave power about 5 mW, and the temperature 4.2°K. This experimental sensitivity begins to approach the theoretical attainable maximum, if the available noise figures for the crystals are substituted in equation (2.31).

This particular 4-mm-wavelength spectrometer has been described and analysed in some considerable detail, since it was specifically designed and constructed for measurements on haemoglobin crystals and similar biochemical compounds, and represents

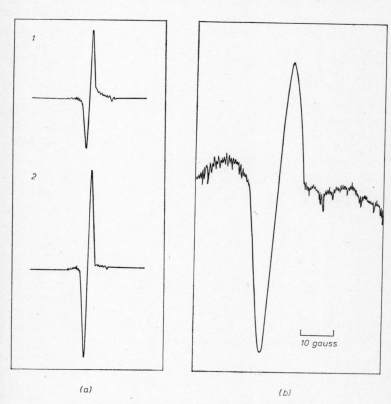

(a)   (b)

Fig. 3.21. E.S.R. signals from 4-mm spectrometer

(a) Comparison of signals obtained from same specimen when using (1) a normal crystal detector and (2) an indium antimonide detector

(b) Signal obtained from 4×10^{12} spins present in a phosphorus-doped silicon crystal

the kind of spectrometer that is ideal for the investigation of very small crystals containing transition group complexes with zero-field splittings. It would appear that many biochemically interesting molecules have these features, and such a spectrometer should be of considerable use in their study.

3.6.3   *2-mm-wavelength spectrometer*

A block diagram of a typical 2-mm-wavelength electron resonance spectrometer is shown in Fig. 3.22, and in this case the microwave power is supplied from the harmonics of a 4-mm klystron,
as used in the 4-mm-wavelength spectrometer described in the last
section. The klystron power supply and automatic frequency-
control circuits are therefore identical with those in Fig. 3.19. The

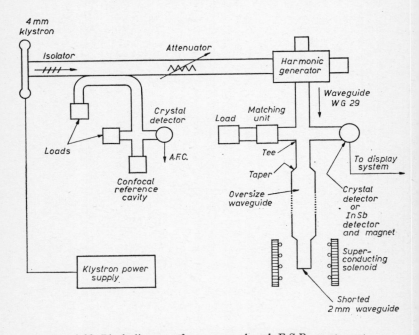

FIG. 3.22. Block diagram of 2-mm-wavelength E.S.R. spectrometer
The very high-frequency microwaves are obtained by harmonic generation from the 4-mm klystron, as shown. Alternatively, harmonic
generation from a higher power 8-mm klystron can be used

4-mm-wavelength power enters a harmonic generator formed by a
silicon crystal mounted across a junction between the 4-mm-
wavelength guide and the 2-mm-wavelength guide (size WG 29).
This waveguide, which acts as a filter to cut out any of the fundamental 4-mm-wavelength power entering from the harmonic
generator, conveys the 2-mm-wavelength power to a magic-T, in
which the third arm containing the specimen is balanced by the
second arm in a normal fashion. The crystal detector, or indium
antimonide detector, is placed at the end of the fourth arm and

feeds the detected absorption signal to the amplifying and recording system as usual.

As in the case of the 4-mm-wavelength spectrometer, it is also
found advisable to use an oversize waveguide to couple the power
from the magic-T to the specimen, which is placed in the superconducting solenoid. Since cavity resonators are extremely difficult
to fabricate at this wavelength, it is normally best to use a length
of shorted waveguide as the reflecting system. The absence of the
higher Q cavity and the lower power available from harmonic
generation reduce the sensitivity of this spectrometer to a value
considerably less than that of the 4-mm-wavelength spectrometer
described in the last section, but the instrument can nevertheless
be extremely useful if large zero-field splittings are to be studied
in transition group complexes.

§3.7

SATURATION EFFECTS

There are two other types of experimental procedure that are often
helpful in the study of biochemical systems and which should be
described briefly as a conclusion to this chapter. One is the determination of the spin-lattice relaxation time, which can give important information on the interaction of the unpaired electrons in the
specimen with the rest of the molecular structure. The other is the
possibility of double resonance experiments on the sample, in
which electron resonance and nuclear resonance are performed at
the same time and coupling between the two is studied. Both
these procedures depend basically on the phenomenon known as
power saturation of the electron resonance absorption, and this
will therefore be outlined briefly before the two techniques themselves are considered.

The phenomenon of power saturation can be best approached
by considering the specific example of the absorption of energy
between two spin levels, such as A and B of Fig. 3.23. Let N_{1_0} and
N_{2_0} be the respective populations of these two energy levels. In
the absence of any input microwave radiation, their ratio will be
given by the normal Maxwell–Boltzman distribution governing
thermal equilibrium at the temperature T_L of the lattice or general
molecular surroundings. Thus

$$\frac{N_{1_0}}{N_{2_0}} = \exp - \left(\frac{\hbar\omega}{kT_L}\right) \qquad (3.1)$$

When the microwave radiation is applied at the resonance frequency, spins in the lower level, B, will be raised to the upper level.

Even although there also occurs the stimulated emission down to level B of those already in level A, there will be a net upward motion of the spins if N_2 is greater than N_1. It is, of course, this net upward transition which actually produces the observed electron resonance absorption. However, it is evident that if this is the only type of transition taking place between the two levels (i.e. a combination of the microwave absorption and the stimulated emission), there will be a steady reduction in the value of N_2 and an increase in N_1 until these reach equality. When this occurs, no further microwave absorption will take place, and hence the actual electron resonance signal will disappear. It therefore follows that there must be some other mechanism which will return the spins

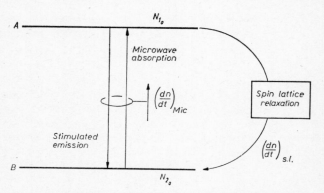

FIG. 3.23. Factors affecting energy level populations

The interactions of the electrons with the incoming microwaves on the one hand, and the lattice vibrations on the other, are represented schematically in this diagram

from the excited level A to the lower level B, but without emitting microwave quanta in the process. This mechanism is the spin-lattice interaction, which allows the spins in the excited level to give their energy to the general molecular, or lattice, vibrations of the system as a whole, and thus return to the ground state by a radiationless transition. This mechanism is represented symbolically on the right of Fig. 3.23, and the transitions produced by it can be written:

$$\left(\frac{dn}{dt}\right)_{SL} \tag{3.2}$$

where $n = N_2 - N_1$. In the same way, the rate of transitions induced by the microwave field can be written as $(dn/dt)_{mic}$, and the total rate of change of the population of the two levels can thus be written as:

$$\left(\frac{dn}{dt}\right)_{\text{total}} = \left(\frac{dn}{dt}\right)_{\text{mic}} + \left(\frac{dn}{dt}\right)_{SL} \qquad (3.3)$$

The first of these terms on the right has, in fact, already been considered before, and is given effectively by equation (2.25) multiplied by the population difference between the two levels. Thus

$$\left(\frac{dn}{dt}\right)_{\text{mic}} = -\frac{\pi}{4} \cdot \gamma^2 \cdot H_1{}^2 \cdot g(\omega - \omega_0) \cdot n \qquad (3.4)$$

The negative sign has been included in this expression to indicate that the effect of the microwave transitions is to reduce the difference in population between the two levels, rather than to increase it. The second term on the right in equation (3.3) can be written as

$$\left(\frac{dn}{dt}\right)_{SL} = \frac{n_0 - n}{T_1} \qquad (3.5)$$

Here T_1 represents the spin–lattice relaxation time, as discussed in §1.3.2, and effectively measures the rate of recovery of the population difference from any disturbing influence, which has altered its value from the normal, n_0, to a different value, n. Under the normal conditions of approximation, as discussed in the paragraphs before equation (2.21), the value of n_0 can be approximated by equating the exponential to the first two terms of its binomial expansion to give

$$n_0 = N_{2_0} - N_{1_0} = \frac{\hbar\omega_0}{kT_L} \cdot \frac{N_0}{2} \qquad (3.6)$$

where N_0 is equal to the total number of unpaired spins in the specimen.

When the new equilibrium conditions have set in, under both the thermal vibrations of the lattice and the presence of the incoming microwave field, the value for $(dn/dt)_{\text{total}}$ must be equal to zero. The new condition of equilibrium will therefore be governed by the equation

$$-\tfrac{1}{4} \cdot \pi \cdot \gamma^2 \cdot H_1{}^2 \cdot g(\omega - \omega_0)\, n + \frac{n_0 - n}{T_1} = 0$$

$$n = n_0[1 + \tfrac{1}{4} \cdot \pi \cdot \gamma_1{}^2 \cdot H_1{}^2 \cdot g(\omega - \omega_0) \cdot T_1]^{-1} \qquad (3.7)$$

The different factors which govern the new equilibrium condition can be related to the actual electron resonance absorption line by obtaining two expressions for the microwave power absorbed. Thus each microwave quantum absorbed corresponds to an energy of $\hbar\omega$, and the rate of power absorption is, therefore, given by

$$P_{abs} = -\hbar\omega . \left(\frac{dn}{dt}\right)_{mic} \qquad (3.8)$$

Equations (3.4) and (3.7) can now be substituted into equation (3.8) to give

$$P_{abs} = \tfrac{1}{2}\,\omega . \omega_0 \left(\frac{\gamma^2 n_0 \hbar}{2\omega_0}\right) . H_1{}^2 . \left[\frac{\pi . g(\omega - \omega_0)}{1 + \tfrac{1}{4} . \pi . \gamma^2 . H_1{}^2 . T_1 . g(\omega - \omega_0)}\right] \qquad (3.9)$$

It will be recalled from equation (2.26) that the gyromagnetic ratio, γ, is equal to $g\beta/\hbar$ and thus the expression in brackets $(\gamma^2 n_0 \hbar)/(2\omega_0)$ is equal to $(g^2\beta^2 n_0)/(2\hbar\omega_0)$. Substitution for n_0 from equation (3.6) then makes this expression equal to $(g^2\beta^2 N_0)/(4kT)$ and comparison with equation (2.21) shows that it is, in fact, equal to the d.c. susceptibility χ_0. With this substitution equation (3.9) may be compared with equation (2.5), which was derived earlier for the power absorbed, to give

$$\chi'' = \omega_0\,\chi_0 . \left[\frac{\pi . g(\omega - \omega_0)}{1 + \tfrac{1}{4}\pi . g(\omega - \omega_0) . \gamma^2 . H_1{}^2 . T_1}\right] \qquad (3.10)$$

It was seen, when discussing the nature of the line-shape function $g(\omega - \omega_0)$ in §1.3.2 and equation (1.9), that a parameter T_2 could be defined as equal to $\pi . g(\omega - \omega_0)_{max}$, and this would also be equal to the inverse of the line-width parameter $\Delta\omega$.

With this substitution, equation (3.10) can be written for the resonance frequency, ω_0, when $g(\omega - \omega_0)$ will have its maximum value, as

$$\chi''_{\omega_0} = \chi_0 \left(\frac{\omega_0}{\Delta\omega}\right) . \left[\frac{1}{1 + \tfrac{1}{4} . \gamma^2 . H_1{}^2 . T_1 . T_2}\right] \qquad (3.11)$$

This equation may be compared with equation (2.29) which was derived for conditions of no saturation. It is now seen that the effective susceptibility, and power absorbed, will be reduced by saturation effects, and equation (3.11) makes clear what actual experimental parameters will produce a noticeable saturation of the resonance line. Thus a large value of the microwave magnetic field strength H_1 will tend to produce saturation, since the spins will then be thrown up to the excited level faster than they can be returned by the spin–lattice relaxation process. Similarly, large values of T_1, which indicate a weak spin–lattice interaction, will also tend to increase the saturation condition, and, in the same way, a large value for T_2, which measures the inverse of the line-width, increases the saturation, since it corresponds to a very narrow line in which the power cannot be distributed widely.

The phenomenon of power saturation can be regarded as having both good and bad effects, so far as the study of electron resonance itself is concerned. On the side of the positive advantages are the facts that it can be used to measure the spin–lattice relaxation time T_1, as is explained in the next section, and that its existence allows changes to be made in the population distribution. These changes can themselves be used to both produce and detect interaction effects at other frequencies; hence they are the basis of all double-resonance experiments, including those in the maser and laser field. These double-resonance experiments are also touched on in the last section of this chapter.

On the side of the drawbacks produced by power saturation may be listed first the inability to use high microwave-power levels, if the spin–lattice interaction time is long, and this means that the sensitivity of the spectrometer will be limited and cannot be increased by increasing the microwave power. Secondly, saturation will produce a broadening effect in the resonance lines, since the saturation will occur in the centre of the line, where more power is absorbed, before it starts to occur in the wings. This can be seen in detail by referring to equation (3.10), where the line-shape function $g(\omega - \omega_0)$ is multiplied by H_1 in the denominator, but not in the numerator, and hence shows that the line will change shape as H_1 itself increases. It, therefore, follows that the onset of power saturation will produce a general broadening of the line as well as a reduction in its expected intensity, and this, of course, reduces resolution as well as sensitivity. Even so, this particular fact can be put to some positive use, since the above analysis of the line-shape broadening only applies to cases of 'homogeneous broadening', where the line is initially broadened by interactions within the spin system or by an external interaction which is fluctuating rapidly in comparison with the time taken for a spin transition. Examples of such interactions are the normal dipole spin–spin interaction between electrons, the spin–lattice interaction itself, and motional and exchange narrowing that can also occur.

This analysis does not apply to external interactions, which have a period of time variation long compared with the time of a spin transition. Such interactions, which may arise from such features as unresolved hyperfine structure or from inhomogeneities in the magnetic field, will, in effect, produce a large number of separate individual absorption lines all overlapping one another. The observed line is thus an envelope of all the unresolved components beneath it and, when power saturation takes place, it will take place for the individual lines separately. Each of these component lines will, therefore, have its height reduced by the same saturation factor, with the net result that the envelope of them all retains its shape, since all parts of it decrease by the same factor. Thus these inhomogeneously broadened lines do not change their shape or width on saturation, and this fact itself can be used to identify them and thus to reveal the probable presence of such things as unresolved hyperfine structure. A detailed description of how saturation effects can be used in such studies may be found in the texts listed at the end of the chapter.[8]

$$§3.8$$

MEASUREMENT OF RELAXATION TIMES

It will be seen in later chapters of this book that considerable information of biological importance can sometimes be deduced from measurements of the relaxation times of the particular molecules being investigated. It, therefore, seems appropriate to give a brief review of experimental methods available for measuring such relaxation times. All these methods of measurement are effectively based on the phenomenon of saturation, as described in the previous section and as represented in equation (3.11). It will be seen from this that the value of T_1 may be derived if the actual value of the saturation factor, Z, can be measured for a given set of conditions for which the magnitude of H_1 and T_2 are already known.

The simplest and the most direct way in which T_1 can be determined, therefore, is to undertake a series of electron resonance absorption measurements on the specimen for successively increasing magnitudes of microwave power in the cavity, and thus successively increasing values of H_1^2. The absolute value of H_1^2 is somewhat difficult to calculate accurately from the microwave circuit constants, and so it is normally determined from measurements on a standard sample, placed in the same cavity with the same microwave power flowing through it. Alternatively, a non-saturating sample can be placed in the same cavity at the same time as the measurements are made. If no appreciable saturation is present, the power absorbed in the electron resonance transition will increase directly with H_1^2, and thus directly with the microwave power flowing into the cavity. Hence the onset of saturation can be determined by the deviation from a linear relation when the electron resonance absorption is plotted against the input microwave power. The variation in the microwave power can be very readily produced, and accurately measured, by employing a precisely calibrated attenuator in the main waveguide run. The only other parameter that then needs to be determined is the value of T_2, and it will be remembered that this is directly related to the line-width of the unsaturated resonance absorption. It can, in fact, be written as

$$T_2 = \frac{1}{\Delta\omega} = \frac{\hbar}{g\beta} \cdot \frac{1}{\Delta H} = \frac{1}{\gamma \cdot \Delta H_{\frac{1}{2}}} \qquad (3.12)$$

In this equation $\Delta H_{\frac{1}{2}}$ represents the line-width at half power before

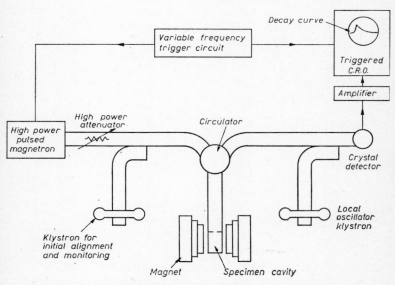

FIG. 3.24. Block diagram of pulse spectrometer to determine
spin–lattice relaxation times
Microwave magnetic fields up to 50 gauss in magnitude
can be produced in such a spectrometer

any saturation has set in, i.e. whilst the relation between the absorption and the microwave power is still in the linear region.

It is evident that such a method of measuring relaxation times does not require any additional equipment to that already provided by an E.S.R. spectrometer. If any wide variation of the spin–lattice relaxation time, T_1, is to be determined, however, it may be necessary to have microwave sources which can give much larger output powers than normal. In fact, if most values of T_1 are to be accurately measured, it is usually better to employ some sort of pulse technique rather than the straightforward continuous-wave spectrometer system. High-power microwave pulses can be obtained from magnetron oscillators, as developed for radar purposes, and these can produce very high-power pulses of energy of short duration. The actual recovery of the signal from such a high input pulse can then be displayed on an oscilloscope screen, and the relaxation time determined directly from the rate at which the signal returns to normal. A block diagram of such a pulse spectrometer is given in Fig. 3.24, and the power from the magnetron is fed, via a circulator, to the specimen in the cavity, as shown. Microwave magnetic fields up to 50 gauss can be produced at the specimen from such high-power magnetron pulses, and the magnetization produced within the specimen by such an input pulse can be written as χ_0''.

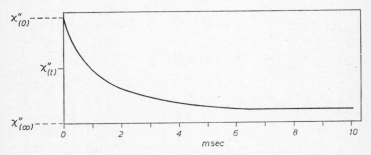

FIG. 3.25. Typical decay curve obtained in pulse studies
The curve shown was obtained from studies by Davis, Strandberg and Kyhl[9] when studying the spin–lattice relaxation time of gadolinium

Immediately after the high-power pulse has been switched off, the strength of the magnetization is monitored by the low-power continuous-wave klystron, which also feeds into the arm of the circulator. The electron resonance absorption that is thus obtained is detected by the superheterodyne system, as shown, and in this way the strength of the signal can be continuously followed, as the magnetization in the specimen returns to its normal value at a rate determined by the spin–lattice relaxation time. Such a decay of the signal is shown in Fig. 3.25, which is taken from the work of Davis, Strandberg and Kyhl,[9] where the spin-lattice relaxation time of gadolinium was being determined. The relaxation time itself can be calculated from the slope of the straight line that is obtained when $\log\left(1 - \chi_t''/\chi_\infty''\right)$ is plotted against time.

These methods of measuring relaxation times by pulse techniques were adapted from earlier methods developed for nuclear resonance spectrometers. Much additional information can be obtained from a complete analysis of such pulse or echo systems, and further details may be found in the references at the end of the chapter. It may also be noted that it is possible to design a spectrometer which can be used to measure both the spin–lattice and the spin–spin relaxation times directly. Such a spectrometer employs two high-power magnetron sources as illustrated in Fig. 3.26. The

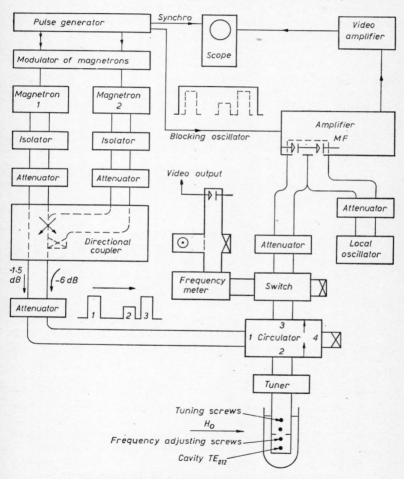

Fig. 3.26. Block diagram of pulse spectrometer employing two high-power microwave sources

This system allows various pulse and echo signals to be obtained and thus accurate measurement of T_2 as well as T_1 (From Kaplan[10])

magnetrons both produce a peak power output of about 200 W, and the pulses are fed to the microwave cavity at 90° and 180° phase intervals. The power passes through an isolator, a directional coupler, two attenuators and a ferrite circulator, before reaching the H_{102} rectangular cavity itself. This cavity is considerably overcoupled to the waveguide through a large diameter iris, and thus has a low Q-value. A free precession signal is then produced by the magnetization of the specimen within the cavity, and this signal passes out of the cavity through the circulator and on to the superheterodyne receiving system, which is used to detect and display it. This spectrometer system is therefore the exact analogue, at a microwave frequency, of the original pulse and echo spectrometers of Hahn,[11] which were developed for nuclear resonance at radio frequencies, and all his detailed analysis can be applied to them. Thus, in this way, the pulse and echo methods previously developed in nuclear resonance can now be applied in electron resonance studies to determine both spin–lattice and spin–spin relaxation times quite accurately.

§3.9

DOUBLE RESONANCE—ENDOR

3.9.1 *Basic principles*

No formal description of nuclear magnetic resonance has so far been given in this book, but the implication has been made several times that this technique is basically the same as electron resonance. The essential difference is that the nuclear magnetic moments instead of the electron moments, are aligned by the applied d.c. magnetic field, and it is these that are being re-orientated by absorption of the incoming electromagnetic radiation. The resonance condition thus becomes

$$hv = g_N \beta_N H \tag{3.13}$$

where β_N is the nuclear magneton, and g_N is the nuclear g-factor, and equal to 5·58 for a proton. Substitution of numerical values in this equation gives the resonance frequency of 42·6 Mc/s for protons in an applied field of 10 000 gauss. Hence the techniques associated with nuclear resonance are those of radio-frequency electronics, and the resonance systems will involve coils and condensors rather than microwave cavities. Thus the whole of the radio-frequency magnetic field, which is again the component which couples to the nuclear magnetic moments, can be concentrated in the inductance coil. The experimental problems are, therefore, normally concentrated on the most efficient way of coupling this coil to the specimen under investigation.

Nuclear magnetic resonance was, in fact, developed a few years before electron resonance, and both were studied independently for some considerable time. The interesting question now arises, however, as to what exactly will happen if both nuclear magnetic resonance and electron resonance are carried out on the same specimen at the same time. There are, in fact, various ways in which such a double-resonance experiment may be conducted. The first

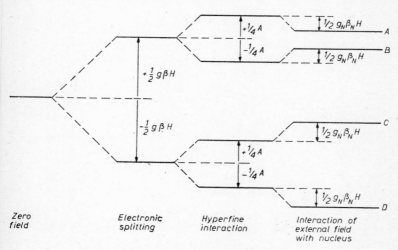

FIG. 3.27. Detailed energy-level splitting associated with proton interaction

The energy-level splittings and shifts associated with the electronic splitting, the hyperfine interaction, and the direct effect of the applied field on the nucleus are illustrated successively. These splittings are not drawn to scale.

of these, due to Overhauser,[12] effectively enables the high sensitivity of the electron resonance technique to be made available in nuclear magnetic resonance studies. Another form of double resonance was suggested a few years later by Feher,[13] who showed that the opposite features could also be attained in double-resonance experiments and that the high resolution associated with nuclear resonance could be obtained in an electron resonance spectrum.

This technique is known as ENDOR (electron nuclear double resonance), and uses high-power microwave radiation to saturate the electron resonance transition under investigation. The applied

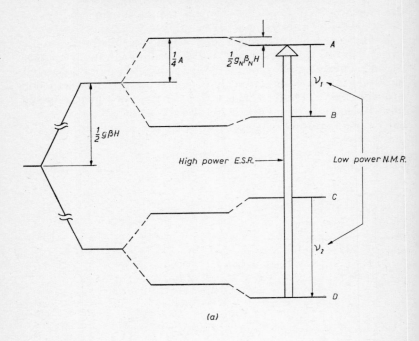

(a)

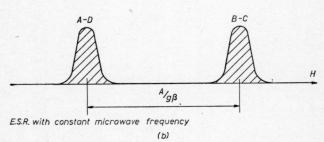

E.S.R. with constant microwave frequency

(b)

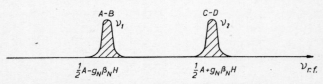

ENDOR with constant microwave frequency and magnetic field

(c)

d.c. magnetic field, and the microwave frequency are held constant, throughout the experiment, at the resonance condition, but at the same time a nuclear resonance frequency is now applied to the specimen. The actual frequency of this additional input radiation is continuously varied, and the effect that it has on the saturated electron resonance line is observed. The principle of the method is probably best illustrated by a simple example, in which a hyper-fine interaction with a nucleus of spin $\frac{1}{2}$, such as a proton, is considered, as illustrated in Fig. 3.27.

Thus each of the original single electron resonance levels are split into two by the interaction with the nuclear spin of the proton, as shown in Fig. 3.27 and this energy of hyperfine interaction can be denoted by $\frac{1}{4}A$. As well as this normal hyperfine interaction between the magnetic moments of the unpaired electron and the protons, there is however a direct effect of the externally applied magnetic field, H, on the nuclear magnetic moment itself. This is usually much smaller than the hyperfine interaction in normal electron resonance spectra, but it will produce a small shift in the four component levels, as indicated in Fig. 3.27. The actual energy of these four levels can therefore now be tabulated as:

Level	Energy
A	$\frac{1}{2}g\beta H + \frac{1}{4}A - \frac{1}{2}g_N\beta_N H$
B	$\frac{1}{2}g\beta H - \frac{1}{4}A + \frac{1}{2}g_N\beta_N H$
C	$-\frac{1}{2}g\beta H + \frac{1}{4}A + \frac{1}{2}g_N\beta_N H$
D	$-\frac{1}{2}g\beta H - \frac{1}{4}A - \frac{1}{2}g_N\beta_N H$

It can be seen from these detailed expressions, and also from the final energy spacing in Fig. 3.27, that the energy difference between levels A and B is now not quite the same as that between C and D.

In the ENDOR technique, the high-power electron resonance frequency is applied to saturate one of the electron resonance transitions, such as that between A and D in Fig. 3.28(a). The result of this saturation will be to increase the population of energy level A, and hence to give level A a higher population than level B. Whilst this saturation is taking place, a radio frequency is also applied to the sample with a frequency such that $h\nu_{\text{r.f.}}$ is equal

Fig. 3.28. (*facing*) ENDOR for single proton interaction
(a) Saturation of E.S.R. absorption and desaturation by radio-frequency resonance
(b) E.S.R. spectrum that is expected
(c) ENDOR spectrum that will be observed

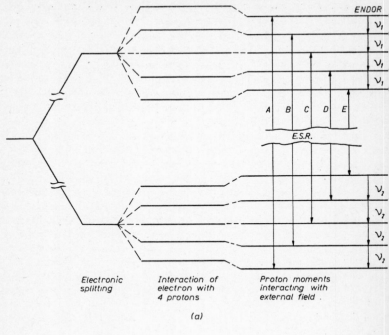

| | | Electronic splitting | | Interaction of electron with 4 protons | | Proton moments interacting with external field . |

Electronic splitting   *Interaction of electron with 4 protons*   *Proton moments interacting with external field .*

(a)

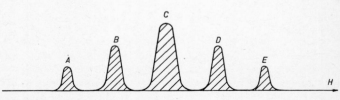

E.S.R. with constant microwave frequency

(b)

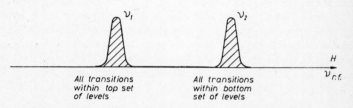

All transitions within top set of levels   *All transitions within bottom set of levels*

ENDOR with constant microwave frequency and magnetic field

(c)

to the splitting between A and B. This then stimulates transitions from A to B and the populations of the two levels will return to their normal equilibrium values. As a result, the saturation of the electron transition will be removed and a strong electron resonance line will then be suddenly obtained in place of the weakened saturation condition. The net result is that, if the detecting system is kept set on the electron resonance signal, a sudden increase in this will be obtained when the nuclear resonance signal sweeps through the condition:

$$h\nu_{\text{r.f.}} = \tfrac{1}{2}A - g_N\beta_N H \tag{3.14}$$

It can also be seen that a similar situation arises when the radio frequency sweeps through the resonance value corresponding to the nuclear transition between levels C and D. The saturation of the electron resonance will have reduced the number of atoms in the level D, but when the nuclear resonance transition is induced by the radio-frequency field the populations of levels C and D will be more or less equalized, and as a result the electron resonance signal being observed will suddenly become desaturated and a large signal will be obtained. It follows from this that, if the radio-frequency signal is swept slowly through a range of values centred on $\nu = A/2h$, a large increase in the electron resonance signal will be obtained when the frequency satisfies either of the conditions given by:

$$h\nu_{\text{r.f.}} = \tfrac{1}{2}A \pm g_N\beta_N H \tag{3.15}$$

From these two values of the radio-frequency resonance signal, the values of both A and g_N can be deduced very accurately.

The normal electron resonance signal that would be expected from such a system is shown in Fig. 3.28(*b*), and compared with the ENDOR spectrum that would be obtained, as shown in Fig. 3.28(*c*). It should be noted that the vertical axis of Fig. 3.28(*c*) represents the actual absorption produced by the electron resonance transition and that both the microwave frequency and the value of the applied magnetic field are held constant throughout the experiment at this resonance condition. The horizontal axis, however, corresponds to the changing frequency of the applied radio-frequency field and absorption lines will therefore be obtained whenever this frequency corresponds to an actual hyper-

FIG. 3.29. (*facing*) ENDOR for interaction with four protons
(*a*) Detailed energy levels for interaction with four equally coupled protons
(*b*) E.S.R. spectrum expected
(*c*) ENDOR spectrum that will be observed

fine splitting present in the overall energy-level pattern. Thus the position of the lines corresponds to splittings and not to different nuclear orientations, and hence no symmetry about a mean point is to be expected.

In fact, if exactly the same analysis had been applied to a system containing four equally coupled protons instead of one, only two ENDOR resonance lines would have been obtained, whereas five separate electron resonance lines are to be expected, as shown in Fig. 1.13(b). In order to explain this, the two sets of five equally spaced component levels, which were predicted in Fig. 1.13 for the four interacting protons, are repeated in Fig. 3.29(a), but also have the additional effect of the external field on the nucleus added to them, as shown. It is evident that although there are still five electron resonance lines, at different frequencies, between these two sets of levels, there will be only two ENDOR transitions of different frequency, i.e. at the frequency corresponding to the energy gap between components of the top set and at the different frequency corresponding to the different gap between the components of the bottom set. The resultant E.S.R. and ENDOR spectra expected in this case are as shown in Figs. 3.29(b) and (c) and it is clear that the ENDOR spectrum is much simpler, and also gives a more direct measurement of the two coupling parameters involved. The difference between the two types of spectrum becomes even more marked if there is unequal coupling to some of the protons, thus one unequally coupled proton causes all the other hyperfine lines to be split by its own interaction. Unequal couplings therefore increase the complexity of the electron resonance spectra by a multiplying factor, whereas each new class of inequivalent nuclei only gives rise to a total addition of two extra lines in the ENDOR spectrum. It therefore follows that any hyperfine pattern in which there are any large number of interacting nuclei, and especially if they are not all equally coupled, will normally produce a hyperfine pattern which is extremely difficult to resolve by ordinary electron resonance means. On the other hand, the ENDOR spectrum remains relatively simple and just possesses two lines for each type of inequivalent proton.

Examples of the way in which ENDOR can be used to analyse various hyperfine interactions in more complicated molecules are given in Chapter 7, where its specific applications to biochemical compounds are considered. Reference forward to Figs. 7.4 and 7.5 illustrates this point for the case of triphenyl-methyl and its derivatives. It will be seen in these figures that the electron resonance spectra themselves are very complex, whereas the ENDOR spectra are relatively simple, and the hyperfine interactions can be deduced from them directly. The ENDOR technique was initially used to study unresolved hyperfine structure in inorganic crystals and proved extremely powerful in resolving splittings which were completely lost in hyperfine patterns which were buried under the normal line-width of the electron resonance spectrum. It has more recently, however, been applied in increasing measure to free radicals in solution, as illustrated by the above example, and in this way promises to be a powerful tool in the elucidation of complex spectra from biochemical and biological specimens. Various examples of its use in this connection are to be found in the following chapters.

3.9.2 *Experimental techniques in ENDOR*

It has already been noticed that the main experimental problems associated with straightforward nuclear magnetic resonance are concentrated on the design of the inductance coil and its coupling to the specimen under investigation. The rest of the spectrometer design is fairly standard electronic circuitry, in which balanced radio-frequency bridges are used, or crossed transmitting and receiving coils are employed, to give a signal when resonance occurs. If a double resonance experiment is to be conducted on any given specimen, then again, the main problems arise in the actual coupling of the radio frequency and the microwave radiation to the specimen in question. Once this problem has been solved, the two spectrometers can effectively be separated to take the normal form of single spectrometers by themselves. The only points which need additional comment here are therefore those on the actual design of the resonance systems into which the specimen can be placed, and which can apply both microwave and radio-frequency fields to it at the same time.

This problem has already been met and partially answered when the question of applying high-frequency magnetic-field modulation to the specimen was encountered. In discussing the ways in which this could be achieved, the use of modulation loops wound around thin portions of the metal cavity was described, and these have been illustrated in Fig. 3.2. It is, therefore, clear that these same kinds of loop could also be used to apply the nuclear resonance radio frequency if it was in this frequency range. The nuclear resonance frequencies can vary quite appreciably, however, and in general it is better to design a resonance cavity system in which radio frequencies reaching up into the megacycle region can be applied at the same time as the microwave power. This calls for cavity walls that are extremely thin, or alternatively, for the insertion of a radio-frequency coil into the cavity itself. Another approach to this problem in which the nuclear resonance coil is wound outside the metal microwave cavity itself is shown in Fig. 3.30, which is taken from a paper by Holton and Bloom.[14] They wound the nuclear resonance coils around the rectangular H_{101} cavity and used a Mylar spacer to hold the two halves of the cavity apart, this having been cut along the broad face. The radio-frequency input then enters the cavity through the slit while the

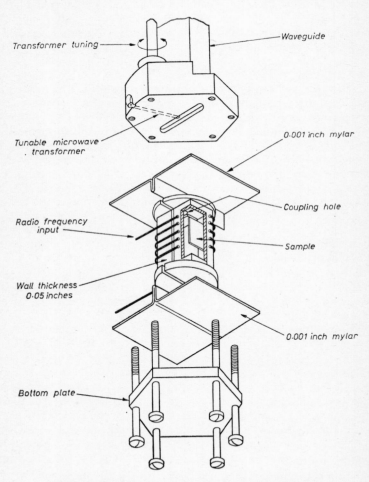

Fig. 3.30. Double resonance cavity employing split walls

A vertical slit down the cavity walls is held apart by the Mylar spacers, and the radio-frequency power produced by the external coil enters the cavity via this slit (From Holton and Bloom[14])

Mylar spaces at the top and bottom of the cavity isolate it electrically at the nuclear resonance frequency. The inductive radio-frequency impedance to the cavity then permits the radio-frequency current to flow on the inner surface of the cavity wall and thus be present at the specimen. It will be seen that this particular design retains fairly thick cavity walls, which have the advan-

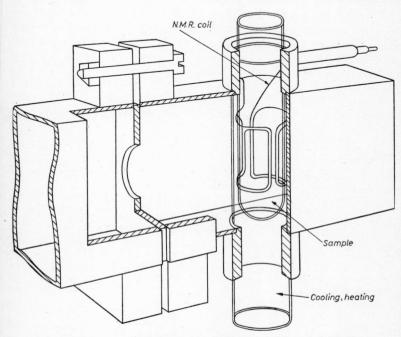

N.M.R. coil

Sample

Cooling, heating

Fig. 3.31. Simple ENDOR cavity with heating and cooling facilities
The N.M.R. coil is mounted inside the cavity in a similar way to internal modulation coils, and a quartz tube carries a stream of nitrogen gas past the specimen (From Kramer, Muller and Schindler[16])

tage of reducing microphonics, and the radio-frequency energy is introduced into the cavity by the techniques outlined above.

A large number of different types of cavities, in which both the radio-frequency input signal and the modulation frequency can be applied at the same time, have been reported in the literature and full details of most of them may be found in the references at the end of the chapter.[15] In the case of biological and biochemical studies, however, a fairly simple cavity design will normally suffice and, since the samples under study often take the form of solutions or viscous material, a cavity system through which a quartz tube may be passed directly has considerable advantages. A very straightforward modification of the normal rectangular cavity can in fact easily be made, as is illustrated in the Fig. 3.31, which shows a straightforward double resonance cavity as designed by Kramer, Muller and Schindler[16] and in which the specimen can be cooled, or heated by appropriate streams of nitrogen gas. Such a cavity introduces very little additional complexity into the microwave circuit, and the only major addition to the normal electron resonance spectrometer is the nuclear resonance frequency required to desaturate the electronic transitions. It is therefore likely that ENDOR techniques may become more widely used in future years, and it is also highly probable that several commercial firms will be offering ENDOR spectrometers on the market as additional accessories to the normal electron resonance systems.

References

1. Wilmshurst, T. H., *Electron Spin Resonance Spectrometers* (Adam Hilger, London, 1966; Plenum, New York).
   Ingram, D. J. E., *Spectroscopy at Radio and Microwave Frequencies*, 2nd ed. (Butterworths, London, 1967).
   Poole, C. P., *Electron Spin Resonance* (Interscience, New York, 1967).
2. Klein, M. P., and Barton, G. W., *Rev. Sci. Instrum.*, 1963, **34**, 754.
3. Yamazaki, I., Mason, H. S., and Piette, L., *J. Biol. Chem.*, 1960, **235**, 2444.
4. Bray, R. C., *Biochem. J.*, 1961, **81**, 189.
5. Bray, R. C., Palmer, G., and Beinert, H., *J. Biol. Chem.*, 1964, **239**, 2667.
6. Mock, J. B., *Rev. Sci. Instrum.*, 1960, **31**, 551.
7. Bogle, G. S., Symmons, H. F., Burgess, V. R., and Sierins, J. V., *Proc. Phys. Soc.*, 1961, **77**, 561.
8. Pake, G. E., Paramagnetic Resonance, Chaps. 6 and 7 (Benjamin, New York).
   Poole, C. P., *op. cit.*, Chap. 18.
   Ingram, D. J. E., *op. cit.*, Chap. 12.
9. Davis, C. F., Strandberg, M. W. P., and Kyhl, R. L., *Phys. Rev.*, 1958, **111**, 1268.
10. Kaplan, D. E., *J. Phys.*, Radium Suppl. No. 3, 1962, **23**, 21A.
11. Hahn, E. L., *Phys. Rev.*, 1949, **76**, 461; 1950, **77**, 297; and 1950, **80**, 580.
12. Overhauser, A. W., *Phys. Rev.*, 1953, **92**, 411.
13. Feher, G., *Phys. Rev.*, 1956, **103**, 500, 834.
14. Holton, W. C., and Bloom, Y., *Phys. Rev.*, 1962, **125**, 89.
15. Wilmshurst, T. H., *op. cit.*, Chap. 4.
    Poole, C. P., *op. cit.*, Chap. 8.
16. Kramer, K. D., Muller-Warmuth, W., and Schindler, J., *J. Chem. Phys.*, 1965, **43**, 31.

Chapter 4

Free Radical
and
Irradiation Studies

GENERAL FEATURES OF A FREE-RADICAL SPECTRUM

It has already been seen that the compounds studied by electron spin resonance can be very broadly divided into two main groups. The first group contains those in which the unpaired electrons are moving in highly delocalized molecular orbitals, and thus embracing several atoms within their wave-functions. The second group contains those in which the unpaired electron is localized much more closely to one single atom, or vacancy site, and interacts quite strongly with the orbital momentum associated with that centre. Most of the first group of compounds can be described by the general term 'free radicals', and these are usually derived from a broken valency bond. Such molecular structures often contain aromatic carbon rings. The unpaired electron associated with the free radical is liberated into a molecular π orbital which can cover most of the atoms in the molecule. There are several general features of E.S.R. spectra which result from such a delocalized electron orbital and two of these are directly concerned with the very small interaction that takes place between the spin and the orbital momentum in such a delocalized orbital.

There are two important parameters in electron resonance studies which both depend directly on the strength of the spin–orbit interaction. The first is the g-value, or, more precisely, its difference from the free spin value of 2·0023. The second is the strength of spin–lattice interaction since the most effective mechanisms for this act via the spin orbit coupling. It therefore follows that most g-values associated with free-radical spectra will be very close to the free-spin magnitude, and, as a corollary, precise information can usually not be obtained from their detailed study. The weak spin–lattice interaction implies that there will be no broadening due to the short lifetime of the spin state. On the other hand, saturation broadening may well take place, since a long spin–

lattice relaxation time, which is associated with the weak spin–lattice interaction, is one of the main causes of saturation effects, as indicated in equation (3.11).

The third general parameter, which is also affected by the highly delocalized orbital, is the hyperfine pattern which is often observed. A brief analysis of this has already been given in §1.5, and it was seen there that a simple pattern of $(2I+1)$ equally intense hyperfine lines is obtained if the unpaired electron interacts with only one nucleus of spin I. The situation becomes more complicated, however, once the unpaired electron is interacting with several different nuclei at the same time. The particular case of an equal interaction with a small number of similar nuclei, such as the protons of hydrogen groups, was also considered in some detail in that section, and it was seen that a typical 'Christmas tree' pattern is obtained. Thus n equally coupled protons will give a total of $(n+1)$ hyperfine lines, with a binomial distribution in their intensities. The more detailed analysis of such complex hyperfine patterns is considered in the next section, but here it may be noted that although the patterns are more complex, it also follows that very specific quantitative information can often be obtained from them concerning the actual distribution of the wave-function of the unpaired electron across the molecular species. This can then often be identified with the active groups in the molecular structure itself, and hence be of very considerable importance when analysing biochemical or biological activity.

Before discussing specific features of the free-radical spectra, it may be wise to consider the difference between spectra obtained from compounds in solution and those obtained from the same molecules arranged in an orderly array in a crystal. In general both the g-value and the hyperfine interaction will vary with the angle between the direction of the applied magnetic field and that of the axis of the internal crystalline electric field, or other symmetry-determining element. Different resonance-field values for the absorption lines and different splittings between the electronic or hyperfine components will then be obtained as the crystal is rotated in the applied magnetic field. Detailed analysis of these angular variations can give very specific information on the orbitals in which the unpaired electron is moving, and hence often on the particular type of chemical bonding associated with the active group in question.

In a large number of cases, however, and especially those concerned with biological and biochemical applications, the system to be studied is essentially in a liquid state and a possibility of studying single crystals does not arise. The individual molecules in the liquid state will, of course, be randomly orientated with respect to one another and with respect to the direction of the applied magnetic field. At first sight it might, therefore, be supposed that a random averaging of the different resonance positions for the absorption line, and the different splittings associated with it, would produce a completely smeared out spectrum, spread over such a wide field range as to be totally unobservable. Fortunately, however, this is in general not the case, and the random orientation associated with the liquid state produces a much more useful form of averaging.

Thus, hyperfine splitting, taken as a specific example, may be shown to consist basically of two different types of interaction. The first of these can be viewed as the straightforward classical case of the dipole–dipole magnetic interaction between the magnetic moment of the electron spin and that of the nuclear magnetic moment concerned. This direct dipole–dipole interaction will have an angular variation given by $(3 \cos^2 \theta - 1)$ and it will be seen that this varies from a maximum of 2 to a minimum of -1 as the angle is varied from 0 to $\pi/2$. The second kind of interaction, which has no direct classical analogy, is known as the contact, or Fermi, interaction and arises from the fact that the '*s*' wave-function of the electron has a finite probability of existence at the actual site of the nucleus itself. Thus, if the electron has any '*s*' character to its wave-function, this will produce a direct interaction with a magnetic moment of the nucleus which is independent of angle θ. These two different types of interaction can be represented by the equation

Hyperfine interaction energy

$$= g_e \cdot g_N \cdot \beta \cdot \beta_N \left[\frac{(3 \cos^2 \theta - 1) \cdot S \cdot I}{r^3} - \frac{8\pi}{3} S \cdot I \cdot \delta(r) \right] \quad (4.1)$$

In this equation g_e and g_N are the *g*-factors of the electron and nucleus respectively, and β and β_N are the Bohr and nuclear magnetons. The first term in the brackets represents the dipole–dipole interaction discussed above; and the angular variation with θ is clearly seen. This interaction will be present from each nucleus close to the site of the unpaired electron, and in the solid state a sum of all such interactions would have to be taken. In a liquid, however, all the molecules are moving rapidly, so that the molecules close to any specific unpaired electron change position rapidly, and usually in a time much shorter than that of the electron-spin transition. It follows that the actual incremental magnetic field, experienced by the particular unpaired electron will, in effect, be an average of all those produced by the different molecules which have been instantaneously present during the course of the spin transition. An average of $(3 \cos^2 \theta - 1)$, taken over all possible angles will, in fact, average to zero. Hence, it follows that, provided the molecular tumbling motion in the liquid is sufficiently rapid, all the unpaired electrons will experience a dipolar field which has been averaged to zero by such motion.

The second term in equation (4.1) represents the Fermi or contact interaction, however, which is quite independent of the angle between the applied magnetic field and the particular molecular orientation, and hence will be the same for each molecule in the liquid state. The isotropic interaction will, therefore, be observed fairly readily from free radicals in liquid solvents, and since most molecular orbitals have some 's' type character, a fairly large splitting, due to this mechanism, is often produced. It should also be noted that the molecular tumbling motion of the liquid state not only averages the anisotropic hyperfine interaction to zero, but also averages the ordinary dipole–dipole broadening between neighbouring electron spins to zero, for exactly the same reasons. In this way the line-width of the resonance absorption can also be drastically reduced by this motional narrowing effect, and this explains why very high resolution spectra can be obtained from the liquid state, when only broad lines are observable from the same compound in a solid. It therefore follows that, if an unknown free-radical system is to be investigated, it is often wisest to study it first in the liquid state, so that only the isotropic hyperfine interactions are observed and measured. Then, once these have been characterized, and if it is also possible to study the same molecules in a single crystal, much more detailed information can be obtained from the anisotropy of the splittings, which can then be observed. It has proved possible to carry out such single crystal investigations in one or two cases of compounds of biological importance, but it is probably fair to say in general that most biochemical and biological studies have been made in solution. Thus most of the information available from these studies has come from the isotropic well-resolved hyperfine patterns that are obtained from the motionally narrowed spectra observed in solutions.

§4.2

THE ANALYSIS OF HYPERFINE PATTERNS

The way in which hyperfine splittings originate has been discussed in some detail, both in §1.5 and in the preceding §4.1. In particular it has been noticed, as illustrated in Fig. 1.11, that a hyperfine pattern consisting of $(2I+1)$ equally intense lines will be obtained if the unpaired electron is interacting with only one nucleus of spin I. The case of interaction with more than one nucleus was also discussed in §1.5. The particular example of one unpaired electron interacting with n equally coupled protons was considered there, and it was seen that, in this case, a hyperfine pattern of $(n+1)$ lines is obtained, with a distribution of intensity across these lines governed by the terms of a binomial expansion.

In a large number of cases, however, and especially those associated with compounds of biochemical or biological interest, the molecular orbital of the unpaired electron will have different coupling coefficients with different protons around the molecular system. In these cases, the hyperfine pattern which is observed may become quite complicated, but it can nevertheless often be unambiguously interpreted, since the very high resolution obtainable from free radicals in the solution allows detailed fitting of the lines in the pattern. A specific example of this may help to indicate the way in which such an analysis may be carried out. Fig. 4.1(*a*) shows the structural formula for the perinaphthene molecule, which consists of three carbon aromatic rings fused together with protons around the edge. In the free-radical derivative of this molecule, one unpaired electron is liberated into the π-bond system, which is formed by the overlapping p orbitals of the carbon atoms, and it can thus move around all the conjugated carbon rings. Fig. 4.1(*b*) shows the electron resonance spectrum that is observed from a solution of this perinaphthene radical[1] and the grouping of the various component lines which form the spectrum is given in Fig. 4.1(*c*) below. It is evident that the spectrum consists of seven main groups of lines which have an approximate intensity ratio of $1 : 6 : 15 : 20 : 15 : 6 : 1$, and that each of these lines is further split into a quadruplet. In order to explain this hyperfine pattern it must, therefore, be assumed that the unpaired electron has a strong interaction with six symmetrically located protons, and this interaction will then produce the seven main components. The further splitting of each of these lines into four sub-components then indicates that there is a further, but weaker, interaction with three other symmetrically situated protons. Consideration of the structural formula in Fig. 4.1(*a*) shows that there is indeed a group of six protons, all identically related within the molecule and labelled *a* in the figure. There is also a group of three protons, labelled *b*, similarly related to each other, but differently from those in group *a*. The ratio of the observed hyperfine splittings, i.e. the 7·3 gauss for group *a* protons, and the 2·2 gauss for group *b* protons, must therefore also measure the ratio of the unpaired electron spin density on these two different types of proton. These experimentally determined unpaired spin densities can, in fact, be compared quite specifically with the spin densities predicted by molecular orbital theory, which in fact predicts the spin densities of 0·325 for each of the protons in group *a*, and of 0·098 for each of the protons in group *b*.

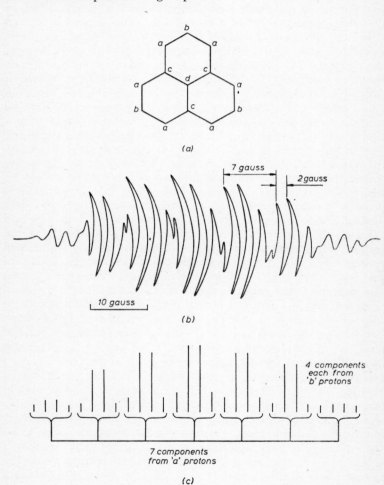

Fig. 4.1. Hyperfine pattern from perinaphthene
(*a*) Structural formula of perinaphthene molecule
(*b*) Observed electron resonance spectrum from perinaphthene radical
(*c*) Grouping of hyperfine lines to show the two different interactions

It is seen from this simple example that such hyperfine patterns can be used to map out the orbital of the unpaired electron in considerable quantitative detail. Moreover, the activity of the different groups within the free radical is often directly related to the concentration of unpaired electrons present, and thus such a hyperfine analysis can also often give direct information on the specific activity associated with different parts of the molecule.

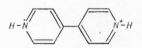

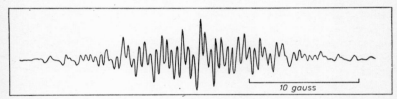

(a)

(b)

FIG. 4.2. Computer synthesis of hyperfine patterns
(a) E.S.R. spectrum observed from bipyridal radical
(b) Spectrum as synthesized by an electronic computer
(From Johnston et al.[3])

The hyperfine patterns observed from free radicals in solution can often be very much more complicated than the example given for perinaphthene. In step with these more detailed experimental studies, very sophisticated theoretical analysis now replaces the original simple theory due to Hückel and others.[2] Since very high resolution is available experimentally, and it is also possible to carry out detailed calculations in this type of theoretical chemistry, very precise comparison can now be made between the observed spectra and the theoretical predictions. In this connection, computers can often be employed to synthesize the predicted spectrum from postulated splitting constants and line shapes. A good example of this use of electronic computers is shown in Fig. 4.2. Fig. 4.2(a) shows the electron resonance spectrum which is actually observed from the bipyridyl radical cation in solution,[3] while Fig. 4.2(b) shows the spectrum which was drawn out by a data plotter, fed from an electronic computer which had been programmed to synthesize the spectrum and produced as good a correlation as possible between that observed and its own predictions. The splitting parameters and electron spin densities used in the production of the best match can then be automatically read out from the computer, and precise details on the molecular orbital of the unpaired electron thus obtained.

Such information can not only give accurate data on the actual orbit of the unpaired electron and the activity associated with different parts of the radical, but the quantitative magnitudes of the splittings can also be used to indicate the types of chemical bonding that are taking place within the molecule itself. Thus, in order to explain this isotropic spectrum at all, it must be assumed that a certain amount of the unpaired electron density reaches the '*s*' orbits of the protons, since electron density confined to '*p*' orbits would produce an anisotropic spectrum, which would be averaged out by the molecular tumbling motion in solution. In the first approximation, the unpaired electrons are, however, only moving in the π orbitals of the ring system, and the '*p*' orbits of a carbon atom, which form the π-bond system, do not directly overlap, or interact with, the '*s*' orbits on the protons. This apparent anomaly was explained by the introduction of what is termed 'configurational interaction', in which a small amount of the higher excited states of the molecule are mixed in to its ground state. These higher states correspond to those in which one electron, of a pair normally held together in a σ bond between the carbon and the hydrogen, is raised to an anti-bonding σ orbital.[4] It is then possible for the one electron left in the bonding σ orbital to be unpaired, and thus interact directly with the proton, since the σ bond is formed from the '*s*' orbit of the attached proton. Detailed theoretical calculations have been carried out on the mechanism of this type of configurational interaction, and the experimentally observed parameters of the electron resonance spectra can now be used to give precise information on the amount of the higher electronic states which are mixed into the ground states of these free radicals. Although this kind of information is probably of more interest to theoretical chemists than to biochemists or biologists, it can, nevertheless, be of considerable relevance when the activity of different radicals is being discussed. It also correlates with studies on the excitation to triplet states, which are considered later in this chapter, and which can themselves be of considerable biological importance.

It is probably fair to say, however, that, up to the moment, the main application of hyperfine pattern analysis in biochemical and biological systems has been as a form of identification of the species present, and, since this is so, further detailed theoretical consideration of the energy levels involved will not be pursued; examples of the use of hyperfine pattern analysis will become apparent in the specific cases considered later.

§4.3

RADICALS FORMED BY HIGH ENERGY IRRADIATION

4.3.1 *Early experimental observations*

It is probably true to say that for the biochemist and biologist there is more general interest in free radicals which exist as part of a normal metabolic process than in those formed somewhat artificially by different forms of irradiation, although a study and understanding of the latter is of increasing importance in such applications as radio-therapy. Further sections of this and the next chapter will be considering the dynamic and transient concentrations of free radicals associated with metabolic processes or enzyme activity. However, since the first investigation of biological materials by electron resonance was, in fact, the early study of free radicals formed by high energy irradiation, this might be the most appropriate way of introducing the study of specifically biological material.

The first studies on biological material by electron resonance were carried out by Gordy, Ard and Shields,[5] in 1955, when they X-irradiated various amino acids and proteins and observed the resulting electron resonance spectra which were produced. Some of their early spectra are given in Fig. 4.3, and in each case the substance studied was in the form of a solid at room temperature, and hence the free radicals formed by the X-irradiation remained trapped in the solid matrix and could be studied at leisure after the irradiation had ceased. In this connection, it should also be noted that the line-widths observed are very large compared with those discussed in the last section, since there is no molecular tumbling motion to average out the anisotropic contributions when radical spectra are being studied in the solid state.

The spectrum shown in Fig. 4.3(*a*) is that obtained when raw silk is irradiated with X-rays, and it is evident that, in this case, a simple spectrum consisting of a single doublet is observed. This

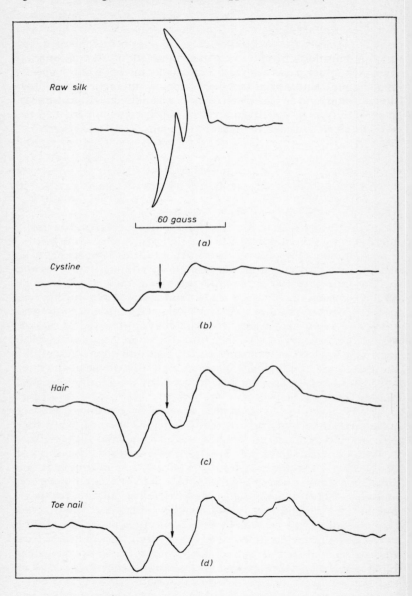

Fig. 4.3. Early E.S.R. spectra from X-irradiated biological specimens
(From Gordy, Ard and Shields[5])

spectrum was, moreover, traced out at two quite different frequencies, i.e. at 9 Gc/s and 23 Gc/s. In both cases, exactly the same doublet with the same splitting of 12 gauss was obtained, indicating that this splitting is indeed a hyperfine interaction and not a spread of g-values. Such a doublet, with two equally intense lines, suggests immediately that the unpaired electron is interacting with one single proton, and in this connection it was noticed that an almost identical spectrum was observed from irradiated glycylglycine. In their initial analysis Gordy, Ard and Shields[5] therefore suggested that this resonance was produced by protons associated with the hydrogen bonds of the polypeptide chains in these molecules.

In contrast to this simple doublet, symmetrically placed about the free-spin g-value, the spectra observed from irradiated cystine and various keratin proteins are shown in Figs. 4.3(b), (c) and (d). It is evident that all these spectra have almost an identical form, and, moreover, this shape is very asymmetrical relative to the free-spin g-value, the position of which is indicated by the arrow. There must, therefore, be some shift in the g-value associated with the free radical giving rise to these spectra, and since such a shift is most unusual in normal organic free radicals, some very specific explanation for this must be found. It will be remembered that the g-values obtained in free-radical spectra are nearly always very close to the free-spin value because the unpaired electron, which is moving in a highly delocalized molecular orbital, does not interact appreciably with the orbital motion of any one atom. There are, however, two atoms which have an electronic structure that can produce quite a noticeable g-value shift, if the unpaired electron interacts appreciably with them. These two atoms are oxygen and sulphur, and if the unpaired electron of the free radical embraces either of these two in its molecular orbital, g-value shifts to as high a value as 2·05 can be produced.[6] Gordy, Ard and Shields[5] therefore suggested that the most likely explanation for this asymmetrical pattern observed from cystine and the keratin proteins was that the unpaired electrons and vacancies which had been liberated by the X-irradiation had, in fact, migrated to the sulphur atoms associated with the polypeptide chain structure and become localized on these atoms.

Thus, when an electron is ejected from a bond in the molecule, the vacancy shifts to the sulphur atom where it has the lowest possible energy, and this then allows the sulphur atom to form a three-electron bond to the second sulphur atom to which it is already linked by the normal electron-pair bond. In this way, the sulphur–sulphur bond might be strengthened, rather than weakened, when it donates an unpaired electron to the original group affected by the irradiation. This rapid transfer of electrons and vacancies down the polypeptide chains, to produce these sulphur–sulphur links, might then act as a protection mechanism against radiation damage; in this way, the sulphur atoms would act as a kind of electron reservoir to repair the vacancies caused by the incident X-irradiation. The fact that exactly the same spectra are observed in X-irradiated hair and nail suggests that the damage produced in these proteins is basically the same as that produced in the cystine molecule itself, and the same explanation in terms of a migration to the sulphur atoms can be used to explain the appearance of these spectra.

These particular observations have been quoted as examples of the early studies of biological material by electron resonance techniques, and the analysis and explanation of the spectra were, of course, highly tentative at the time. More recent work has altered some aspects of this interpretation, but these early experiments laid the basis for detailed investigations which followed, and initiated studies on the exact way in which the sulphur atoms could act as a form of protection mechanism for radiation damage.

4.3.2   *More recent studies*

More recent investigations on the electron resonance spectra that are produced when proteins and amino acids are irradiated[7] have sought to elucidate the actual mode of formation of the free radicals, and to follow both the primary and secondary products of the initial irradiation damage itself. Gordy's original work[5] was generally confirmed in that it became clear that, when proteins are irradiated at room temperature *in vacuo*, two main types of electron resonance are observed. On the one hand, a well-resolved doublet, as previously discussed in the case of silk, and on the other hand, a broad resonance spread over approximately 100 gauss when observed at X-band wavelengths and corresponding to the spectrum attributed to the sulphur atoms in the last section.

In order to check on this last point in more detail, Gordy and Kurita[7] then studied single crystals of cystine dihydrochloride that had been γ-irradiated, so that the variation of the *g*-value could be measured precisely. Fig. 4.4 shows that in an arbitrary direction two doublets were observed, but these are found to coalesce along the crystallographic '*b*' axis and in the '*ac*' plane. This suggests that both doublets arise from the same free radical, but that these are differently orientated with respect to the crystalline axes. It is also clear from Fig. 4.4, that the splitting within each doublet is independent of microwave frequency, and is thus produced by a hyperfine interaction, whereas the separation between the two is frequency dependent, being noticeably greater at 23 Gc/s than at 9 Gc/s. This shows that considerable *g*-value shifts must be associated with these particular resonance lines.

The actual resonance field positions of the two components of these two doublets are plotted out in Fig. 4.5, for measurements

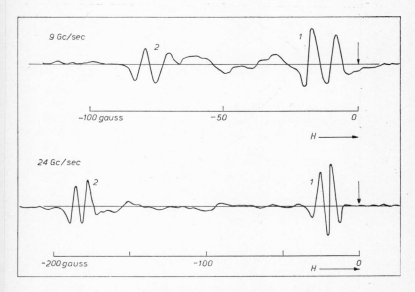

FIG. 4.4. E.S.R. spectrum from γ-irradiated cystine dihydrochloride

The spectra are plotted for an orientation where the applied magnetic field makes an angle of 30° to the *b* axis, and hence both doublets are seen

The top spectrum is obtained at 9 Gc/s, while the lower spectrum is at 24 Gc/s, and it is evident that the splitting between the components of the doublets is independent of frequency, indicating a hyperfine interaction, while the separation between the two doublets increases linearly with frequency, corresponding to a *g*-value difference (From Kurita and Gordy[7])

in the '*bc*' plane at 9 Gc/s, and relative to the free-spin *g*-value as zero. The experimentally measured points are indicated by the open circles, and these can then be fitted to curves calculated for different *g*-value variations. From these, and similar measurements in the other crystallographic planes Gordy and Kurita[7] deduced that the three principal *g*-values for these radicals were $g_x = 2 \cdot 003$; $g_y = 2 \cdot 029$ and $g_z = 2 \cdot 052$.

The large anisotropy in this *g*-value variation can then be explained by assuming that the incident γ-radiation breaks the S—S

K

bond of the cystine molecule, and releases an unpaired electron which remains highly localized on the S atom. This interpretation is supported by the fact that two differently orientated radicals are produced, while all the original molecules are aligned in equivalent orientations. Thus, if the observed resonances were due to a three-

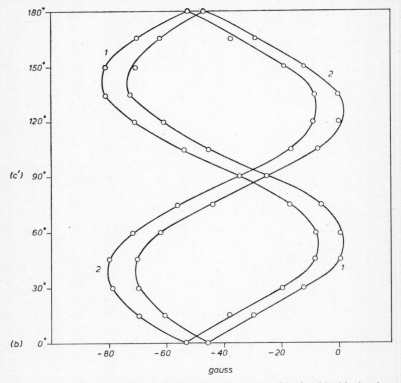

FIG. 4.5. E.S.R. spectrum from γ-irradiated cystine dihydrochloride in the *bc* plane

The field positions of the components of the two doublets are shown relative to a constant *g*-value of 2·0036. The calculated curves are drawn as the solid lines to give a best fit with the experimentally measured points, represented by the open circles (From Kurita and Gordy[7])

electron bond of the S—S link, as previously postulated, only one type of free-radical orientation could be expected. Breakage of this bond allows the molecular fragments to reorientate in two different ways, however.

This interpretation is further confirmed by the hyperfine structure analysis. Thus, to establish whether the doublet splitting was

due to a hydrogen bonded to a carbon atom, or from one bound to
an oxygen or nitrogen atom, comparison was made with spectra
obtained from deuterated samples; since the D atoms will exchange
with the protons bound to oxygen or nitrogen, but not with those
bound to carbon atoms. No discernible change in the hyperfine
pattern was obtained, and it was, therefore, concluded that the
interaction must be with a proton bound to a carbon atom. It
would, therefore, seem that the wave-function of the unpaired

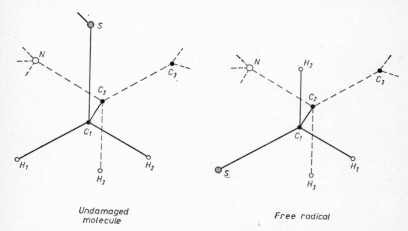

Undamaged
molecule

Free radical

FIG. 4.6. Reorientation of free radicals formed by γ-irradiation of
cystine dihydrochloride

The structural formula of the reorientated free radical is shown at
the bottom of the figure and can be compared with the structural
formula of the undamaged molecule shown above. Breakage of the sul-
phur-sulphur bond allows the sulphur atom to swing down to the
left, as shown, with a corresponding rotation in the positions of H_1
and H_2 (From Kurita and Gordy[7])

electron, which is mainly localized on the sulphur atom, overlaps
with one of the protons of the adjacent CH_2 group, by a process of
hyperconjugation, and thus produces the observed doublet splitting
of about 10 gauss. The resultant orientation of the free radical
will therefore be as indicated in Fig. 4.6.

The analysis of these particular results has been considered in
some detail, since they show how much more information can be
obtained from single-crystal studies, if the crystals are available,
and they also afforded conclusive evidence on the origin of the
large g-value spread found in so many irradiated proteins.

Once this g-value variation is known in a single crystal, it is then
possible to predict the line shape that would be obtained from a

polycrystalline sample containing free radicals of this material, but orientated completely at random with respect to the applied magnetic field. This theoretical calculation was, in fact, carried through in a general way by Kneubuhl,[8] and if the results of his analysis are applied to the *g*-value variation of the cystine dihydrochloride crystals, an average for the centre of the resonance is found to be 2·028. These experimental measurements on a single crystal, and their extension to the polycrystalline case does, therefore, confirm that the main feature of the free radicals formed by *γ*-irradiation of sulphur-containing proteins is a localization of the unpaired electron on the sulphur atoms, but probably not in a three electron S—S bond as initially postulated.

At the same time, a considerable amount of work has also been carried out to obtain further clarification on the nature of the doublet hyperfine splitting observed when dipeptides and polypeptides are X-irradiated. In an analysis of their more recent measurements, Gordy and his collaborators[9] suggest that this doublet may be derived from an unpaired electron which is localized near one proton in a site along the protein back-bone, instead of from an interaction with a proton linking across in an OH bond, as earlier suggested. It is always difficult to make an unambiguous interpretation of the electron resonance spectrum when polycrystalline or amorphous material is being studied, and, as seen for the case of the cystine dihydrochloride, much more specific and definite information can always be obtained from studies of a single crystal.

Thus, single-crystal measurements had already been made on malonic acid,[10] succinic acid,[11] glycine,[12, 13] *d-l*-alanine,[14] and glutamic acids. Gordy and his colleagues therefore proceeded to make systematic studies on as many single crystals of proteins and their amino acid constituents as possible. A comparison of these spectra and a theoretical prediction of how these would appear when averaged over all possible directions suggest that the observed doublets are produced by free radicals in the back-bone of the polypeptide chains, and that these are probably formed in all of the amino acid residues when a side chain is broken off. Work by Henriksen *et al.*[15] has suggested that the probable location of the unpaired electron is, in fact, in the glycine residue, since only one hydrogen atom has to be removed in this case.

As a general summary of these studies, it can, therefore, be said that two types of radical are eventually formed when proteins are X-irradiated. If the protein contains sulphur atoms, then radicals of the type I shown below are produced, with the characteristic *g*-value spread associated with the sulphur interaction. The other type of radical, producing the doublet, is probably normally produced by radicals of type II, where there is a direct interaction of the proton with an unpaired electron located on the protein backbone. On the other hand, studies of polyamino acids[16] which have been X-irradiated, suggest that two other types of radical, type III and IV below, are sometimes produced.

Type I

$$\begin{array}{ccc} H & H & O \\ | & | & \| \\ ...N & -C & -C... \\ & | \\ & H-C-H \\ & | \\ & S\cdot \end{array}$$

Type II

$$\begin{array}{ccc} H & & O \\ | & \cdot & \| \\ ...N & -C & -C... \\ & | \\ & H \end{array}$$

Type III

$$\begin{array}{ccc} H & & O \\ | & \cdot & \| \\ ...N & -C & -C... \\ & | \\ & H-C-H \\ & | \\ & H \end{array}$$

Type IV

$$\begin{array}{ccc} H & & O \\ | & \cdot & \| \\ ...N & -C & -C... \\ & | \\ & H-C-H \\ & | \\ & X \end{array}$$

It would also seem that radicals of type III are also produced on X-irradiation of the proteins themselves, although definite evidence for radicals of type IV has only been obtained from the constituent amino-acids so far. The detailed investigations of Patten and Gordy[9] on this topic can be summarized in Table 4.1 where the different amino acids, or proteins, are listed on the left, and their composition is then given. The different types of radical spectra, as indicated by the four types above and obtained on X-irradiation, are indicated in the right-hand columns.

The form of spectrum observed from the type III radical is illustrated in Fig. 4.7(*b*) in comparison with the more common doublet spectrum of Fig. 4.7(*a*). It will be seen that there is likely to be an equal interaction of the unpaired electron with the three protons of the methyl group in the type III spectra, and hence a quartet hyperfine splitting is expected, as observed for the irradiated poly-l-alanine of Fig. 4.7(*b*).

In order to check on the postulated origin of these free-radical signals and those observed experimentally from the proteins, a direct comparison was also made between the signals obtained

TABLE 4.1 Radical types produced by X-irradiation[9]

Protein	Amino acid composition					Radical type			
	Alanine	Glycine	Cystine	Cysteine	Other residues	I	II	III	IV
Polyglycine		100					x		
PolyAlanine	100							x	
PolyLeucine					Leucine				x
PolyGlutamic acid					Glutamic acid				x
PolyCystine			100			x			
Gelatin	8·0	23	0	0			x	x	
Collagen	8·2	23	0	0			x	x	
PolyAspartic acid					Aspartic acid				x
Histone	68	5	0	0			x	x	
Casein	2·9	1·8	0·3	0	Leu 14 % Glutamic acid 20 %		x	x	x
Zein	94	0	0·8	0				x	x
Edestine	4·0	0	0·9	0·5		x	x	x	x
Gliadin	2·0	0	0	2·5		x	x	x	x
Pepsin	0	5·8	1·5	0·5		x	x	x	
Ovalbumin	6·3	2·8	0·5	1·2		x	x	x	
Chymotrypsinogen	4·9	5·3	2·8	0·9		x	x	x	
Lysozyme	5·3	5·2	6·2	0		x	x	x	
BovineAlbumin	5·6	1·6	5·1	0·3		x	x	x	
Ribonuclease	0	1·2	6·2	0·6		x	x	x	x

from the proteins themselves and the signals obtained from proportional mixtures of the amino acid constituents of the particular protein being considered. These measurements were undertaken systematically by Patten and Gordy[17] under controlled conditions of temperature and pressure, and a similar set of investigations was

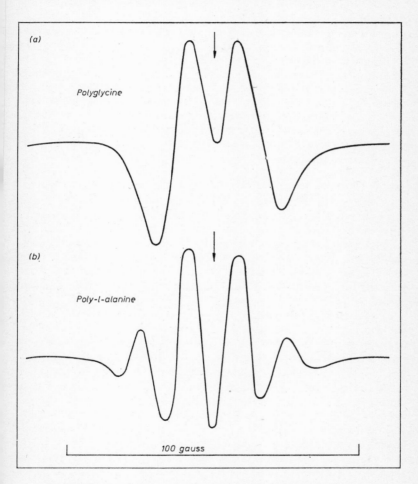

FIG. 4.7. Different types of spectra observed from X-irradiated proteins
(a) From Type II radicals showing the single doublet from the CH interaction
(b) From Type III radicals, showing the quartet hyperfine pattern obtained from an interaction with the CH_3 group
(From Patten and Gordy[9])

also undertaken on fibro-proteins, when mixtures of solid solutions of various constituent amino acids were studied by Pohlit *et al.*[18]

It will be seen that these measurements have now reached a stage when the probable site of the unpaired electron, and free-radical activity, can be identified for the different proteins. Thus, even although these solid-state spectra give broad lines which are difficult to identify and analyse, a significant amount of information can nevertheless be obtained from this type of investigation. These results can be compared with the recent work of Reisz and White[19] who have used tritiated hydrogen sulphide to intercept the free radicals produced under these irradiation conditions and thus determine where the radicals are distributed along the carbon chain. Their results indicate that free-radical sites do appear to be widely distributed among the various amino acid residues of the backbone of the chain, and it would appear that the specific activity of glycine may not be quite so high as that initially suggested by the electron resonance analysis.

4.3.3   *The mechanism of formation of secondary radicals*

The experimental evidence which has been summarized in the previous sections makes it clear that the radicals which are eventually observed by electron resonance after high-energy irradiation of biological tissue are not normally those which are initially produced when the radiation first interacts with the material. The interesting question, therefore, arises as to the nature of the actual mechanism of secondary radical formation from the initial radiation products.

The first suggestions for possible mechanisms were made by Gordy and his co-workers,[17, 20] who proposed a form of intramolecular migration process. Thus they assumed that the first step was the random formation of electron holes throughout the compound by the incoming ionizing radiation. It was then suggested that the electron, or electron hole, migrates within the protein molecule to the sulphur-containing groups and glycine residues, and at these sites the secondary radicals are formed by the release of a hydrogen atom. Such a process requires, of course, the fairly easy and rapid motion of an electron, or hole, along the protein structure, and it was proposed that this might take place by an overlap of wave-functions from one peptide group to another, across the asymmetrical carbon atom. This type of mechanism is illustrated in Fig. 4.8, where the orientation of the peptide planes is indicated and the possible overlap across the intervening carbon atom can be visualized.

A somewhat similar type of mechanism which would also

account for large electron mobility was put forward by Blumenfeld and Kalmanson,[21] since their initial work suggested that there was less radiation damage produced in protein molecules themselves than in their constituent amino acids, when these were given the same dose of radiation. These workers therefore suggested that there must be a form of conduction band system, set up by an interaction across the spiral helix of the polypeptide chains. Such a possible conduction channel is illustrated in Fig. 4.9, where the

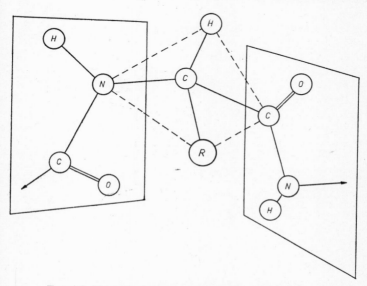

FIG. 4.8. Migration of electron along protein backbone
Mechanism postulated by Gordy,[17,20] in which migration of the electron, or electron hole, takes place from one peptide group to another, across the asymmetrical carbon atom, by hyperconjugation

hydrogen bonds are shown as the linking mechanism whereby the electron wave-functions can overlap between successive spirals of the chains and thus give a direct and rapid path for the movement of electrons down the protein backbone. Further, more detailed work, on the quantitative numbers of damaged centres produced in proteins, as compared with those produced in the constituent amino acids, has suggested, however, that there is really no significant difference between these two cases, and, therefore, the necessity for such a rapid electron conduction channel down the protein molecule no longer exists. It would, therefore, appear that this initial suggestion of an irradiation protection mechanism, involving conduction bands down the proteins, may no longer be required, although its possible existence is an interesting point which future experiments may be able to confirm or eliminate.

As an alternative to these mechanisms, which envisage the formation of secondary radicals by electron migration, Henrikson and his colleagues[15, 22] have suggested that the secondary radical

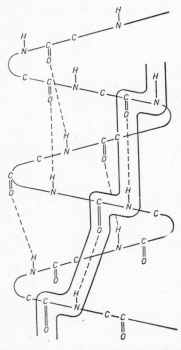

Fig. 4.9. Migration of electrons via conduction band system
Mechanism postulated by Blumenfeld and Kalmanson,[21] in which conduction band systems are set up across the protein structure by the linking mechanism of hydrogen bonds

formation takes place by a direct interaction with molecular fragments, rather than by electron migration itself. In this process, it is assumed that the initial centres, which are not detected by the electron resonance, consist of the mixture of positive and negative ions formed by the ionizing radiation, together with radical fragments, which have been produced by the actual breakage of chemical bonds. It is then assumed that these primary radicals react with each other and with neighbouring unaffected molecules to produce the secondary protein radicals. This type of mechanism can, therefore, be described as an inter-molecular process, rather than an intra-molecular process discussed before. At first sight it might appear to be rather difficult to undertake any experiments to distinguish between these two possible kinds of mechanism. On the other hand, however, if the formation of the secondary radicals really takes place by an actual interaction between molecular fragments and previously intact molecules, rather than by electron migration within a given molecule, it might be possible to demonstrate this by adding certain characteristic molecules into the system and seeing if these themselves were attacked in the process of forming the secondary radicals.

This was the technique used by Henriksen *et al.*, who studied both the primary and secondary radical spectra observed by electron resonance when specific proteins were mixed with smaller molecular compounds and the composite mixtures were then X-irradiated. These mixtures were first irradiated at 77°K, and the spectra were then observed before any heating to room temperature. As might be expected, these spectra showed that both the protein molecules themselves, and the added smaller molecular compounds, had suffered damage and both had electron resonance centres detectable in them. It was also seen from these measurements that the addition of the new substance apparently had no noticeable effect on the primary radicals formed in the protein since these spectra were identical with those obtained on irradiation of the protein alone. These mixtures were then heated to room temperature, however, and thus the primary radicals were allowed to react with each other and with the other molecules present. The electron resonance spectra were then again observed, and it now became obvious that the unpaired electrons initially induced in the proteins had become transferred to the added compounds present in the mixture, and this result was also checked by the use of radioactive sulphur atoms. These experiments can, therefore, be taken as fairly conclusive proof that at any rate some of the secondary radicals are produced by an inter-molecular mechanism, in which molecular fragments actually react together, although they do not eliminate the possibility that some of the secondary radicals might still be formed by an intra-molecular migration of electrons down the protein molecules themselves.

4.3.4 *Correlation of secondary radicals and biological damage*

One of the most interesting questions that is raised by these studies is how much the actual biological damage to a given system is affected by the formation of the secondary radicals. One way of

trying to obtain experimental evidence of this point follows closely
along the lines already outlined in the previous section by the work
of Henriksen *et al.* Thus, in Fig. 4.10, the inter-molecular mechanism of secondary radical production is illustrated diagramatically,
and it is further assumed that the secondary room-temperature
radicals formed may then go on to interact with the biological
function of the molecules and produce such effects as the loss of
enzyme activity. If this is in fact the case, then a direct correlation
between the production of secondary radicals and biological
damage would be expected. An experimental check on the validity
of this proposal was first attempted by Hunt *et al.*,[23] who measured
the number of secondary radicals induced in ribo-nuclease from

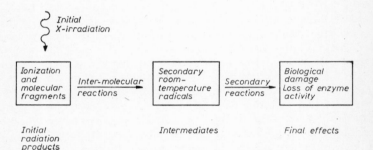

FIG. 4.10. Mechanisms of secondary radical formation
and resultant biological damage

The various steps in the formation of secondary radicals, and their
reactions to produce biological damage, are shown schematically.
Direct correlation between the concentration of secondary radicals and
loss of enzyme activity might thus be expected

the electron resonance spectra observed, and then compared this
with the actual loss of enzymatic activity which could also be
measured. They did, in fact, find a direct correlation between the
number of radicals produced and the inactivation of the enzymatic
activity, and hence concluded that a large fraction of the biological
damage was produced by the secondary radicals. This work was
followed by further studies which employed the effect of oxygen
on the radicals formed, where it was shown that treating the
irradiated specimens with molecular oxygen produced both a
change in electron resonance spectrum and also a change in the
inactivation yield.

This work appears to be further supported by a more detailed
study of the qualitative change in spectra when oxygen treated.
Thus it is found that the doublet type of spectrum is noticeably

affected by the presence of the oxygen, whereas the part of the spectra due to the sulphur-containing groups is not affected. This is probably because the SH groups, and their production, are independent of the presence of oxygen, whereas the doublet spectra produced from the electrons down the protein chain itself, represent sites which are both susceptible to oxygen and capable of deactivating the enzymatic function of the molecule. It should be stressed, however, that these results are of a somewhat tentative nature, but they do show that electron resonance spectra can give a great deal of useful information in studies of the biological damage produced by high-energy irradiation and also on possible mechanisms of radiation protection.

§4.4

RADICALS PRODUCED BY OTHER FORMS OF RADIATION

4.4.1 *Production by thermal hydrogen atoms*

During the last few years, various groups of research workers have been investigating the radicals that are produced in biological materials by thermal hydrogen atoms, and other types of particle or radiation treatment. Thus Ingalls and Wall[24] have been studying the effect of thermal hydrogen atoms on polymers, while Cole and Heller[25] followed this up with studies on malonic acid, thymine and DNA, whilst other groups[26, 27] have been studying their effect on pyrimidine and purine derivatives, and on various amino acids. In general, it can be stated that the effects produced by bombardment with thermal hydrogen atoms are the same as those produced by bombardment with X-radiation. Some differences can, however, be detected, and this has been brought out by some recent studies by Jensen and Henriksen,[28] who produced the thermal hydrogen atoms in a hydrogen gas discharge tube in the normal way, but passed the products through a system of bends before they interacted with the biochemical compounds, and in this way they have been able to separate out the different effects produced by the thermal hydrogen atoms on the one hand and the ultra-violet radiation, which is produced at the same time, on the other. They have thus been able to make a detailed comparison of the radiation-induced free radicals which are produced by (i) X-rays, (ii) ultra-violet radiation and (iii) thermal hydrogen atoms.

In certain specific cases in particular, such as that of thiol-penicillanine, the spectrum produced when the compounds were bombarded by thermal hydrogen atoms was identical with that produced when the compounds had been X-irradiated, and this spectrum had already been accurately identified.[29] It would therefore appear that exactly the same radicals are formed in the two cases, both the spectra having the distributed *g*-value variation typical of the sulphur atom, and that the unpaired spin density is residing mainly in the $3p$ orbital of the sulphur atom. The radicals are therefore probably formed by proton abstraction in both cases, as indicated below

$$\begin{array}{c} CH_3 \\ | \\ HOOC\diagdown \quad\quad\quad | \quad\quad\quad\; \centerdot \\ \quad\quad CH\!-\!C\!-\!SH + \dot{H} \;\rightarrow \\ H_2N\diagup \quad\quad | \\ CH_3 \end{array} \quad\quad \begin{array}{c} CH_3 \\ | \quad \centerdot \\ HOOC\diagdown \quad\quad | \\ \quad\quad CH\!-\!C\!-\!\dot{S} + H_2 \\ H_2N\diagup \quad\quad | \\ CH_3 \end{array}$$

Further work is continuing on these thermal hydrogen atom studies, which promise to provide useful additional information on the radicals formed under these conditions, as compared with those produced during the X-irradiation studies themselves.

4.4.2 *Study of transient radicals formed by high-energy electrons*

In all the work considered so far on the free radicals which are produced by gamma irradiation, X-irradiation or other high-energy sources, the substances studied have been in the solid state. As a result, the free radicals which have been formed have been trapped in the solid matrix and not allowed to react with each other, or other molecular fragments, and thus decay. If the main interest of the irradiation studies is to discover the effect of irradiation on biological tissue as close to its natural environment as possible, rather than to elucidate in detail the nature of the free radicals themselves, then a study of the free radicals formed when solutions are irradiated would be very much closer to the reality of a biological situation.

The great difficulty of studying the formation and interaction of free radicals which are formed by high-energy irradiation of solutions is the fact that they only last for a very small length of time before reacting with other free radicals present, and thus decaying away. There will, of course, always be a dynamic concentration of such transient radicals built up, but if their reaction rates are very fast, this concentration may well be below the limit of detection of the electron resonance spectrometers available. One method of overcoming this disadvantage is to use very high-energy doses of radiation and thus build up large dynamic concentrations of the transient free radicals, when they can be studied in an ordinary electron resonance spectrometer.

One example in which such a dynamic concentration of transient radicals was maintained in solution at room temperature is given by the work of Fessenden and Schuler[30] who produced such radicals by irradiating a solution of liquid ethane with high-energy electrons (Fig. 4.11). These electrons had been accelerated to a voltage of 2·8 MeV and this beam was then directed on to the sample, held in the specimen-holder inside the cavity resonator of the spectrometer. This operation is not so simple as it may appear at first sight, since the electron beam will normally be bent very

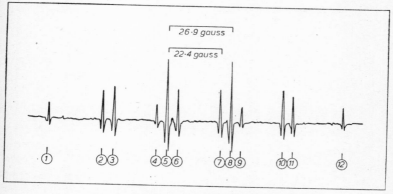

FIG. 4.11. E.S.R. spectrum from liquid ethane
under electron bombardment

The absorption lines here shown are from a dynamic concentration of the free radicals present in the liquid state. The twelve numbered lines are from the ethyl radical, while the weaker ones between lines 5 and 6, and between 7 and 8, are from the methyl radical (From Fessenden and Schuler[30])

severely and defocused by the presence of the magnetic field which is required for the actual resonance experiment itself. To overcome this basic difficulty, Fessenden and Schuler[30] arranged to bring the electron beam into the cavity along the axis of the magnet, by drilling a hole right through the magnet system and bringing the beam down an evacuated pipe, which reached into the cavity and ended at the specimen tube. In this way, the electron beam itself remained focused until it reached the liquid hydrocarbon and very efficient electron irradiation was thus secured.

The kind of spectrum which they obtained is given in Fig. 4.11 which shows the pattern observed from irradiated liquid ethane; it should be stressed that this spectrum is taken from a liquid sample in which the radicals are quite free to interact and disappear, but in which a sufficiently high density of radiation is being maintained to produce the observed dynamic concentration of the radicals. The twelve numbered lines on the spectrum arise from the ethyl radical, while the weaker ones between lines 5 and 6, and 7 and 8, are caused by the methyl radical. The particular interest of these results in so far as their applications are related to biochemical and biological studies, however, are the kinetics which can also be experimentally deduced from such investigations. These kinetics can be followed, both in terms of the build up and decay

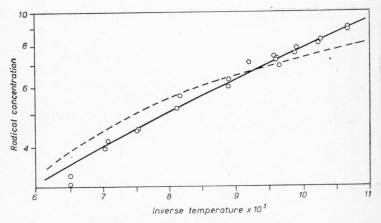

FIG. 4.12. Kinetics of radical recombination

The variation of ethyl radical concentration with temperature of the liquid ethane is plotted as open circles for the experimental points, and compared with two possible theoretical predictions (From Fessenden and Schuler[30])

of the dynamic concentrations with time, and also from the temperature dependence of the radical concentrations. Thus Fig. 4.12 shows the way in which the measured radical concentration varies with the inverse of the temperature of measurement, and compares the experimentally measured results with two sets of theoretical predictions which are based on slightly different assumptions. The correct fitting of the theoretical curve to the experimental results enables both the first- and second-order rate constants for these reactions to be determined and thus gives very direct information on the kinetics of the radical formation and decay in such a system. The limited availability of high-energy electron accelerating machines, and of magnetic systems suitable for use with them, tends to limit such studies at the moment, but it is likely that future work may tend to develop in this field of investigation, where the actual conditions are so much closer to those existing in a biological system than is the case of the solid or frozen specimens studied in earlier work.

§4.5

STUDIES ON PHOTO-SYNTHESIS AND TRIPLET STATE EXCITATION

4.5.1 *E.S.R. studies on photo-synthesis*

The actual mechanism of photo-synthesis has been investigated very intensively by a variety of different techniques over the past few years, and as soon as the potentialities of electron resonance in biochemical and biological studies were appreciated, these techniques were applied to investigate the photo-synthetic process itself. Two groups of research workers made the initial studies in this field, Commoner, Heise and Townsend,[31] on the one hand, and Sogo and his collaborators,[32] on the other. In the initial experiments of Commoner and his colleagues, green leaves were lypholized before insertion into the cavity of the spectrometer, and hence kinetic studies on the specimens were not possible. However, these initial experiments did show that there were significant differences in the concentration of unpaired electrons observed in these specimens when they had been grown in the presence or absence of light, as indicated in Fig. 4.13. This seemed to suggest immediately that there might be some correlation between the unpaired electrons that were present in the specimens and the actual photo-synthetic process itself.

Further experiments were, therefore, undertaken by the two groups of workers, in which it was possible to study aqueous solutions of chloroplasts *in situ* in the cavity under different conditions of illumination. It was for such work as this that the aqueous absorption cells previously discussed were, in fact, first developed. These studies showed that there were only a very small number of unpaired electrons in the absence of any illumination, but that this concentration increased over sixfold as soon as high-intensity light was applied to the specimens. The absorption signals observed had no hyperfine structure and a line-width of about 10 gauss, which was very similar to the width being obtained at that time from other metabolically active tissues.

Kinetic studies on these specimens were then made and the growth rates of the signals were measured by the use of rapid recording techniques. It was found that the unpaired electron

L

concentrations rose exponentially to a steady value after the onset of illumination, with time constants of the order of 12 sec. This again correlated with other kinetic studies on the photo-synthetic process and suggested close correlation between this process and the unpaired electrons being observed. Further studies[33] were also made under a variety of temperature conditions, with the type of cavity illustrated in Fig. 4.14.

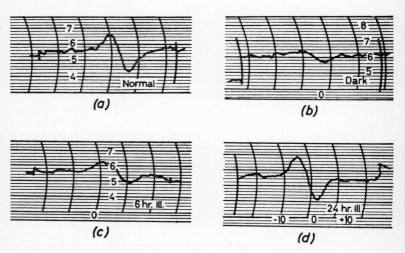

(a)   (b)

(c)   (d)

FIG. 4.13. Unpaired electron concentration and photo-synthetic activity
These spectra show how the unpaired electron concentration observed in barley leaves varies with conditions of illumination
(a) From seedlings grown in normal illumination
(b) From seedlings grown in the dark
(c) Grown in the dark and then illuminated for 6 hours
(d) Grown in the dark then illuminated for 24 hours
(From Commoner *et al.*[31])

As a result of the kinetic studies made with this kind of spectrometer, it was shown that the signal *growth* time appeared to be more or less independent of the temperature of the specimens, even when this varied from liquid nitrogen temperatures to room temperature. On the other hand, the signal *decay* time was very noticeably affected, and normally increased by many orders of magnitude when the temperature was lowered to that of liquid nitrogen. The fact that the signal growth time was effectively temperature independent tends to rule out mechanisms of photo-synthesis in which the unpaired electron concentration is directly associated with either chemical reactions or triplet state excitation. On the other hand, the fact that the decay time is very much increased at low temperatures suggests that the unpaired electrons that are being detected must lie in some form of energy-level traps. The thermal energy associated with room temperature could then relatively easily excite them out of these traps, but such energy would not be available at the lower temperatures.

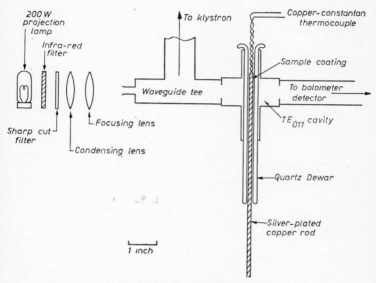

FIG. 4.14. Cavity and spectrometer system
used for initial photo-synthetic studies
The chloroplast solution was painted on the surface of the silver-plated copper rod, which passed through the quartz dewar shield. The temperature was varied by inserting the bottom end of the rod into a suitable temperature bath (From Sogo, Pon and Calvin[32])

Further measurements, at higher resolution,[65] then showed that there were two components in the E.S.R. signal. One, centred on a g-value of 2·002 with no hyperfine structure, was only observed after illumination and had rapid onset and decay rates. The other component, centred on a g-value of 2·005, had five hyperfine peaks, was sometimes present in the dark, and had much longer rise and decay times. These results suggested that the first component was due to the unpaired electrons intimately associated with the photo-synthetic process, while the second component is probably a free radical intermediate associated with the oxidation–reduction process.

These kinds of results and deductions therefore led to the suggestion that the unpaired electron concentrations, observed by electron resonance in these studies on photo-synthesis, were probably associated with a series of low-lying traps spaced a little below the first excited state and might well be in the direct run of the photo-synthetic pathway itself. The complete elucidation of this mechanism does, of course, require a detailed comparison of the electron resonance measurements with the results obtained by other techniques. It is, however, of considerable interest that electron resonance can again afford a very direct probe into one of the basic mechanisms associated with the life process itself.

4.5.2   *Triplet-state excitation in proteins*

The work on photo-synthesis also focused attention on the study of triplet-state excitation in proteins, and a considerable amount of investigation is now taking place in this field. The general features of triplet state excitation and the type of electron resonance signal that will be obtained have already been discussed briefly in §1.4.2. It was pointed out there that the high anisotropy of the electronic splitting will normally cause the $\Delta M_s = \pm 1$ transitions to be smeared out beyond detection, unless single crystals are used. However, the $\Delta M_s = \pm 2$ transition, which is not always completely forbidden, will correspond to the same energy gap whatever the angle between the magnetic field and the molecular field axis. This situation is in fact illustrated in Fig. 4.15, where the constant energy difference for the $\Delta M_s = \pm 2$ transitions is illustrated for a variety of angles. The possibility of detecting this $\Delta M_s = \pm 2$ transition at reasonable intensity has revolutionized the study of triplet states in biochemical and biological specimens, since these are very seldom available in the form of single crystals, and hence any spectrum which is anisotropic will tend to be lost completely. The fact that triplet-state excitation would be detected by electron resonance from these double quantum jumps was first demonstrated in the case of glass-like material, by the work of Van de Waals and de Groot,[34] and triplet-state studies of aromatic amino acids were made by Ptak and Duzou[35] in 1963, and by Shiga and Piette[36] in 1964. These last two workers were also able to detect triplet-state excitation in proteins themselves in 1964, and confirmed the correct interpretation of the signal by its correlation with the phosphorescence which was observed at the same time.

This work has been followed up by a variety of workers, and the investigations of Shiga, Mason and Simo[37] can be quoted as an example. They examined various features of the triplet states, observed in proteins and model compounds, in order to obtain more information on the relation between the triplet state and the ligand system, and the oxidation state of the metals in metallo-proteins. In these experiments, the sample solution was placed in a quartz tube, cooled to liquid nitrogen temperatures, and then

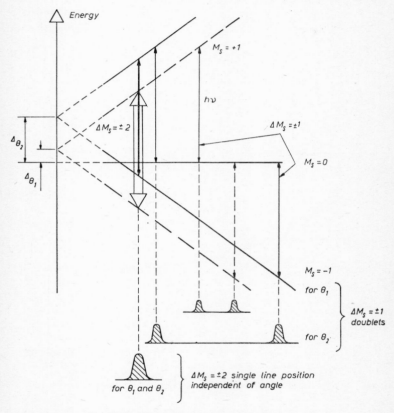

FIG. 4.15. Energy level system for triplet state
The $\Delta M_s = \pm 2$ transition is shown by the arrows on the left and it is evident that this has the same magnitude, whatever the zero field splitting between the $M_s = 0$ and $M_s = \pm 1$ states

inserted into the electron resonance cavity. The samples were ir-radiated *in situ* through a window in the cavity with light from a monochromater, fitted with a 200-W high-pressure mercury lamp. At the same time as the electron resonance signal itself was re-corded, the phosphorescence of the specimen was also measured, and both signals could be fed either to a dual trace oscilloscope or to suitable integrating computers to increase the sensitivity. A typical electron resonance signal, obtained in this way, is shown in Fig. 4.16(*a*), and the decay curve for this signal is shown below in Fig. 4.16(*b*). This signal is actually from the triplet state of glycyl-*l*-tryptophan, in a 1 : 1 glycyl phosphate buffer solution, and it

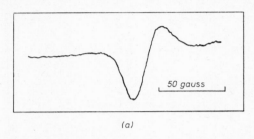

(a)

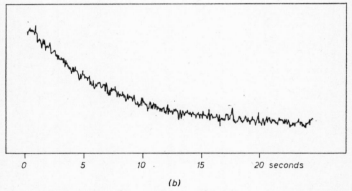

(b)

FIG. 4.16. E.S.R. signal observed from triplet state of glycyl-*l*-tryptophan

    (*a*) Observed first derivative plot of absorption line, obtained from 10^{-3} M solution excited with 290 mμ wavelength at 77°K

    (*b*) Decay curve of triplet excitation when illumination is cut-off (From Shiga, Mason and Simo[37])

is seen to have a half-life of 4·5 sec. The fact that the decay of this electron resonance signal does correspond to the decay of the triplet state itself, can be clearly seen by a direct comparison of the decay curve for the electron resonance absorption, with that of the phosphorescence of the triplet state. Two such curves are shown for comparison in Fig. 4.17, and both are seen to decay in exactly the same fashion, once the illumination has been switched off.

It is clear from this brief review of a single example, that electron resonance is, in fact, a very powerful technique for the study of triplet-state excitation in proteins, and the effect of oxygen and the presence of other metal atoms can be studied in detail. Thus in the particular paper being considered, Shiga, Mason and Simo[37] showed (i) that self-quenching of the tryptophan triplet would occur above a certain critical concentration, (ii) that the nature of the solvent has a variable effect on the triplet state lifetimes of different proteins, which suggests that the actual conformation of the protein is being affected in these cases, (iii) that the presence of

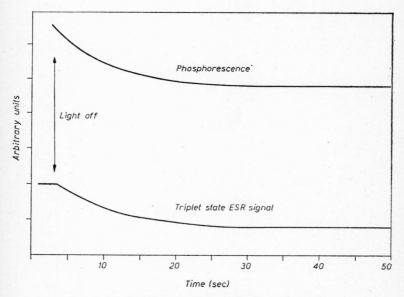

FIG. 4.17. Comparison of decay curves of triplet state
The E.S.R. signal and phosphorescence obtained from the triplet state of tryptophan are compared. Excitation at 290 mμ, with tryptophan in 1:1 glycerol phosphate buffer at 77°K (From Shiga, Mason and Simo[37])

certain paramagnetic ions, such as the cupric ion, can quench the triplet-state excitation completely, and (iv) that oxygen has a very marked effect on the triplet-state yields in haemocyanin, and that oxy-haemocyanin shows smaller triplet-state yields than de-oxy-haemocyanin.

Various other studies on triplet states in different protein molecules are being undertaken by several groups now, and this is obviously a field in which electron resonance will continue to give some very interesting and specific results.

§4.6

RADICALS FORMED BY PYROLYSIS

One of the first electron resonance studies to have direct implications for biochemical and biological systems was associated with
the discovery that quite high concentrations of unpaired electrons

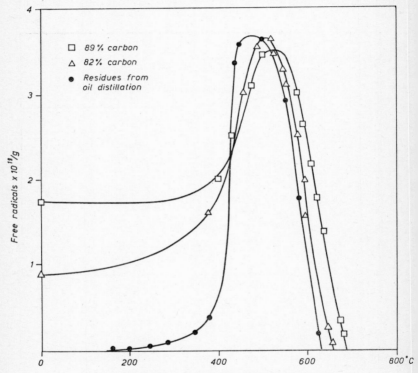

FIG. 4.18. Variation of free-radical concentration
and temperature of carbonization
The variation of free-radical concentration with carbonization temperature for samples with different initial carbon content

became trapped in the condensed carbon ring systems, which are
formed when organic matter is carbonized at relatively low temperatures. The observation of these high concentrations of unpaired electrons was reported at more or less the same time by one
or two different groups of workers,[38, 39] and in each case it was
found that the same general variation of unpaired electron, or free-

radical, concentration with temperature was obtained whatever the starting material. These initial sets of results are summarized in Fig. 4.18, where the unpaired electron concentration observed in the carbonaceous material is plotted against the different temperatures of carbonization. It is evident that there is a steep increase in unpaired electron trapping, or free-radical formation, as the carbonizing temperature approaches 600°C, but that soon after this there is an apparent rapid fall in concentration, and above carbonization temperatures of 800°C, very little electron resonance absorption signal is observed. In the region of carbonization between 1000°C and 1500°C it is in fact very difficult to detect any electron resonance absorption at all, but at temperatures of carbonization above this range, E.S.R. signals reappear.

These early measurements were also correlated with similar signals that could be obtained from various naturally occurring coals and anthracites, and, in this case, the unpaired electron concentration was found to vary in a somewhat similar way, but with coal ranking, or total percentage carbon content,[40] as indicated in Fig. 4.19. It is seen that there is again a rapid increase in unpaired electron concentration as the percentage carbon content rises above 90 per cent, and that a maximum again appears in this variation so that for the very high ranking anthracites the electron concentrations are beginning to drop. As in the case of the carbonized material, this region of falling free-radical concentration was found to be associated with a rapid increase of electrical conductivity in the specimens concerned, and this suggested that small graphitic-type planes were then beginning to be formed in the carbonaceous material. The studies on the different types of coal and anthracite enabled a fairly detailed correlation to be made with X-ray studies that had also been carried out on these compounds.[41] A comparison with these results seemed to suggest that the trapping of the unpaired electrons occurred as the condensed ring system grew in the pyrolysed organic material, as illustrated in Fig. 4.20. A simple picture for this free radical formation could, therefore, be put forward as associated with the thermal breaking down of the normal carbon–hydrogen bonds in the formation of the condensed carbon rings. Every so often in this process, one of the edge bonds around the carbon condensed rings would also be removed, thus leaving a free-radical site and liberating an unpaired electron into the carbon ring system, where it would become highly stabilized by the large molecular orbit available to it in the π-bond system over the ring structure. Some such conjugated ring system was necessary to explain the high stability of these carbon radicals, since measurements on charcoal from some of the Egyptian pyramid tombs indicated that the free-radical concentration was still present at about the expected concentration.

The fairly rapid disappearance of the signal, when electrical conduction began to be significant in these pyrolysed organic materials, could then be explained by the conjugated carbon rings beginning to join in large numbers in a coherent fashion to form the graphitic planes responsible for the conduction, mutually satisfying their unpaired broken bonds in the process. Such

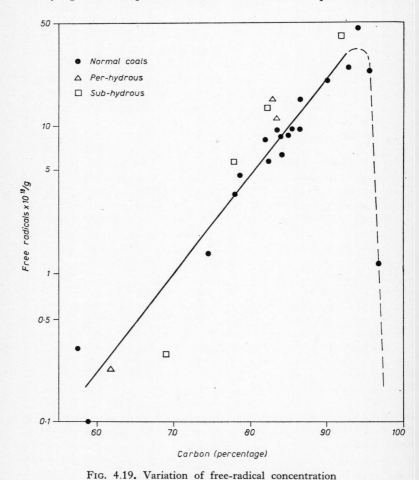

Fig. 4.19. Variation of free-radical concentration
with percentage carbon content

The variation is seen to be of the same general nature for different kinds of coal graphitic planes would then have conduction electrons associated with them, but the electron resonance absorption from these would tend to be broadened very considerably by all the defects present in the structure. Thus any unpaired electrons that were present in the carbonization range of about 800° to 1500°C would probably have their absorption broadened beyond detection. The fact that electron resonance absorption could also be obtained from some

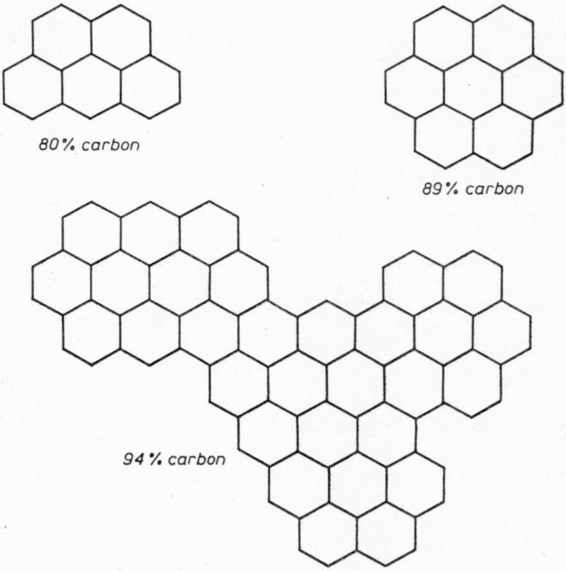

80% carbon

89% carbon

94% carbon

FIG. 4.20. Growth of condensed ring systems

The growth in size of the condensed carbon ring systems as the percentage carbon content rises can be clearly seen. The larger rings offer high stability for an unpaired electron released into them

organic material carbonized at much higher temperatures, and from material which has an obvious graphitic nature associated with it,[42] could be explained by a simple extension of the initial theory. Thus these E.S.R. signals could be interpreted as associated with conduction by charge carriers, in contrast to the lower temperature signals, due to the free radicals localized on the edge of the carbon ring system, or more colloquially termed 'broken bonds'.

Studies on the variation of the line-width of these signals also seemed to confirm this viewpoint. The low-temperature carbons all appeared to have much the same width of about 10 gauss, until the very high percentage carbon contents were reached. Since this did not appear to vary much with concentration of the unpaired electrons, nor significantly with temperature of observation, the width was attributed to unresolved splitting due to surrounding protons, rather than direct interaction with other unpaired electrons, or a spin–lattice effect with the molecule as a whole. These line-widths did change, however, as the graphitic state was approached, again confirming the simple picture of a change from 'broken bonds' to 'conduction-type electrons' previously outlined.

One striking effect was discovered shortly after the initial identification of these radicals, and it became known as the 'oxygen

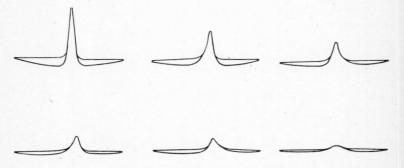

Fig. 4.21. Oxygen effect on carbonized charrs
The decay of the E.S.R. signal as oxygen is admitted to the carbon charred *in vacuo* is shown from left to right across the top and then the bottom. This effect is reversible on evacuation

effect'.[43, 44] It was found that if the carbons were pyrolysed under vacuum or nitrogen, and the unpaired electron concentration then observed, high concentrations could be obtained, but that if oxygen was then admitted to the specimen, the signal intensity would rapidly decrease as the oxygen diffused through the carbonized material. This effect of letting oxygen into such a sample is illustrated in Fig. 4.21, and the rapid decay of the signal can be clearly seen. The very striking feature of this effect was, however, the fact that if the oxygen was then pumped away from the specimen, the process was quite reversible and the large initial signal was re-obtained as the sample was evacuated. This experiment demonstrated very clearly that this effect must be of a physical, rather than a chemical, nature and therefore due to some interaction between the paramagnetic properties of the oxygen molecules and the unpaired electrons in the carbon structure. Further detailed study of this oxygen effect[45] indicated, however, that there were in fact two different mechanisms present, since in a number of cases the initial admission of oxygen would irreversibly reduce the signal, and this could only then be partially restored by pumping the oxygen away. Such studies as these, and associated measurements of the line-width at the same time, seem to indicate there are two processes present. Thus, on the one hand, a physical broadening process must exist which is associated with the absorption of molecular oxygen on the surface and a dipolar interaction between the oxygen molecules and the unpaired electrons. At the same time, however, there is also often present a chemical interaction, in which real pairing of electrons takes place, thus reducing the integrated absorption, and this is not reversible on evacuation.

The detailed elucidation of the exact mechanisms present in those carbonaceous compounds is very difficult owing to the complex and relatively irreproducible nature of the specimens themselves. A very good review paper summarizing the early electron resonance studies on these carbons was produced by Singer,[46] and the advances that have taken place in more recent years can be briefly summarized as follows:

(i) By carbonizing specific organic molecules in solution it has been possible to obtain some well-resolved hyperfine patterns on the E.S.R. signals in the early stages of carbonization, and hence the site of the 'broken bonds', or free radicals, so formed can be identified.[47, 48]

(ii) Recent studies[49] on the high-temperature carbons (above 1500°C) suggest that the signal may contain two components, one due to localized centres (and therefore obeying Curie's law and with a g-value very close to 2·0023) and the other due to conduction electrons (and therefore possessing a Fermi-Dirac variation with temperature and a g-value greater than 2·0023). Mrozowski[49] suggests that these two types of electron may interact by some exchange mechanism to produce a single line with average characteristics. The study of the temperature variation of the signals seems to confirm the existence of these two components and it would appear that neutron bombardment of the specimens increases the proportion of localized broken-bonds present, whilst doping with alkali metals increases the proportion of conduction electrons.

(iii) More detailed studies on the g-values observed in these high-temperature graphitic carbons suggests that there may be a correlation of this with layer spacing and crystallite size on the one hand[50] and the diamagnetic susceptibility on the other.[51]

It rapidly became clear that these studies on pyrolysed organic matter might have considerable applications in quite a variety of directions, such as in the coal and petroleum industries, and to the general subject of catalysis involving carbon, or charcoal, materials. It was also rapidly appreciated that this might well have applications in a strictly biological sense, and that the high unpaired

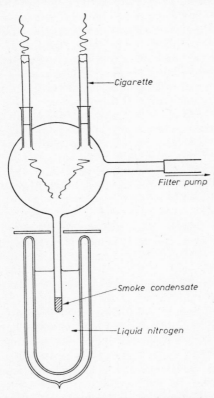

FIG. 4.22. Schematic diagram of 'smoking machine'
The smoke condensate collects in the tail immersed in liquid nitrogen, and this is then transferred to the E.S.R. spectrometer

electron concentrations associated with these condensed rings in carbonaceous matter might well link with such processes as carcinogenic activity, especially since substances such as cigarette smoke and diesel oil fumes would both contain very high quantities of such carbonized material.

Some initial experiments[52] were, therefore, carried out to see if there was any direct correlation between the measured unpaired electron concentration and the carcinogenic activity of such materials as cigarette smoke. One of the early experiments to test this theory is illustrated in Fig. 4.22, which shows a 'smoking machine' designed to smoke four cigarettes at the same time, and operated from a filter pump. This applied intermittent suction to the main chamber, at such a rate and pressure differential as to cause the cigarettes to glow in the same manner as when smoked by a normal human subject. The smoke thus formed was drawn into the spherical chamber below, and a small finger then led down from this into a Dewar of liquid nitrogen, as indicated. Quite a high proportion of the smoke thus formed was condensed into the tip of this cooled finger, and any free radical containing material, thus condensed, was deep frozen and further chemical action thus prevented. The size of the bowl and length of the finger were adjusted so that the time taken for the smoke to reach the deep-frozen position was approximately the same as that for smoke to reach the human lung after inhalation at the mouth.

Sufficient cigarettes were first smoked in the machine to produce a reasonable quantity of deep-frozen condensate in the finger tip. This was then rapidly transferred to the cavity of the electron resonance spectrometer, which had already been pre-cooled, and the total free-radical content of the condensate was then measured. This concentration thus included both the stabilized radicals formed in the normal carbonized organic matter, as described in the previous paragraphs, and also any active radicals which had been formed during the smoking process, but which had not had time to react and disappear in the few seconds of lifetime down through the chamber to the frozen finger. The smoking machine was then removed from the electron resonance cavity and the condensate was warmed to room temperature and kept at this temperature for about 30 minutes. It was then deep-frozen again, replaced in the electron resonance cavity, and the free-radical concentration again measured. This second measurement would then give the concentration of stable radicals present in the condensate, any active ones having had plenty of time to react and thus disappear after warming to room temperature. Such measurements were, of course, not very reproducible, because of the various parameters involved in the design of the smoking machine. The measurements did indicate quite clearly however, that stabilized radicals and active radicals were both present in the original smoke condensate to an appreciable extent.

It is not surprising that the normal stabilized radicals associated with the condensed rings of the pyrolysed organic matter, are present, since this is very general phenomenon. The presence of more active radicals, which disappear on warming to room temperature, indicates, however, that more active agents are also present in the cigarette smoke, and these might well cause active chemical interaction when entering the lung tissue itself. It was, of course, impossible in these preliminary experiments to attempt any identification of such active species, but these experiments did indicate that there may well be a link between unpaired electron concentrations and carcinogenic activity. This point is followed up in the next section where other more sophisticated methods of approach, also employing electron resonance methods to help elucidate the mechanism of carcinogenic activity, are discussed.

§4.7

E.S.R. AND CARCINOGENIC ACTIVITY

In the last section it was seen that high concentrations of unpaired electrons were detected by electron resonance in some substances which were known to be carcinogenically active,[52] and this suggests that the presence of the unpaired electrons might be associated in one way or another with the carcinogenic activity itself. Although these particular experiments did not produce any detailed theory on the mechanism of carcinogenic activity, at the same time as they were being performed, some fairly sophisticated theories, linking carcinogenic activity with the detailed molecular orbital electron configuration of the molecules in question, were being propounded.

This molecular orbital approach to an explanation of the mechanism of carcinogenesis originated with the work of Professor Pullman.[53] The Pullmans applied, in effect, the same theory to the carcinogenic-type molecules that Huckel[2] had previously applied to the simpler molecules discussed earlier in §4.2. They then postulated that the interaction of the carcinogenic aromatic hydrocarbon with the biological cellular material would take place through certain specifically active regions of the hydrocarbon, if the π electrons of the molecule had been allowed to spread out and enter the bonds which unite the hydrocarbon to the cellular material. As a result, these bonds themselves would become partially double, and the whole electronic cloud of the hydrocarbon would then assume an approximately quinonoid configuration. This theory has the advantage of very general application to hydrocarbon molecules, and the experimental results which are available[54] do seem to confirm that there is indeed a quinonoid nature in the bonds between the hydrocarbon and the cellular tissue.

Professor Pullman's more detailed predictions[55] as to which aromatic hydrocarbons would be carcinogenic, and which would not, can probably be best summarized in the diagrammatic form as in Fig. 4.23. In this figure, two different regions of an aromatic molecule are defined, i.e. the K region and the L region. Pullman's theory then states that in order to be carcinogenic, the molecule must possess a reactive K region (a reactive aromatic bond), but it must also be devoid of an active L region (reactive para positions). The reactivity of these two regions can be expressed quantitatively in terms of the 'localization energies' which in turn measure the activation energies for the chemical reactions which can occur at these two types of site. The theory can thus be put on a strict quantitative basis, and the quantitative correlation which has been

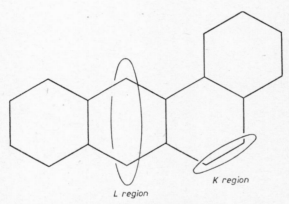

Fig. 4.23. K and L regions of aromatic hydrocarbons

The crucial regions for reactivity, according to Pullman's[55] theory of carcinogenic activity, are known as the K and L regions, as defined in this figure

established is able to explain, with very few exceptions, the carcinogenic activity, or inactivity, of all polybenzenoid hydrocarbons which have been tested experimentally.

The actual threshold 'localization energies' can best be expressed in quantitative forms, in terms of the exchange integral obtained from the molecular orbital calculations, and known as β. Pullman's theory then says that for a hydrocarbon to be carcinogenic, it must possess a K region with an index smaller than $3 \cdot 31\beta$, but that if it possesses an L region in its structure, then it will not be carcinogenic, unless the index associated with this region is greater than $5 \cdot 66\beta$. A typical way in which this theory can be applied to specific molecules is illustrated in Fig. 4.24 for various derivatives of anthracene. Fig. 4.24(*a*) shows the structural formula for anthracene itself with the corresponding activities of the K and L regions, expressed in terms of the exchange integral, as explained above.

It is seen that anthracene does not have a K region with an index smaller than 3·31, and its L region is not greater than 5·66, and therefore on both counts it will not be carcinogenic. The case of naphthacene, which is formed by adding one extra benzene ring linearly to the anthracene molecule, is shown in Fig. 4.24(b). Here again the two indices for the K and L regions are shown and, again,

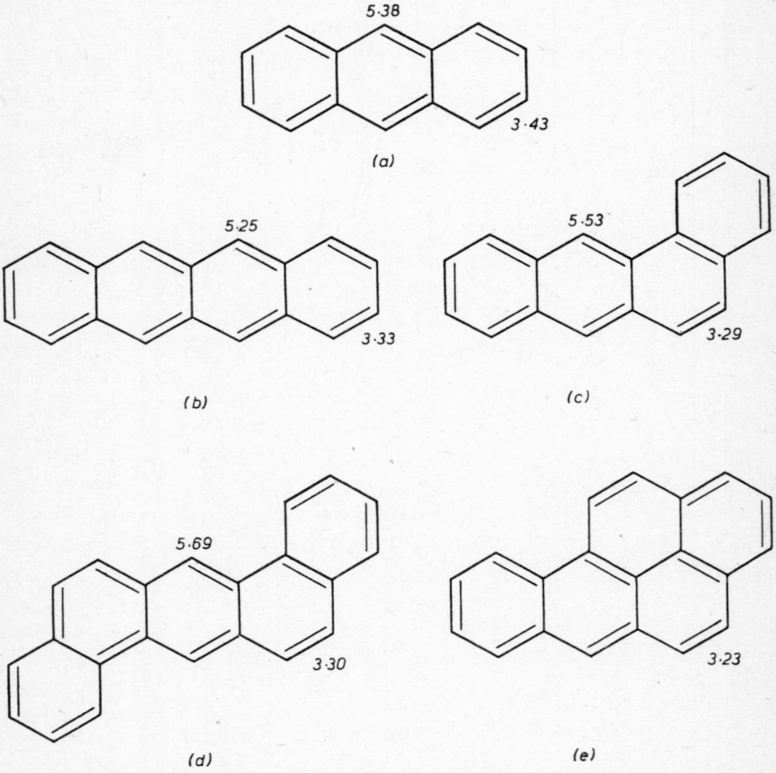

FIG. 4.24. Reactivity indices for different aromatic hydrocarbons

The reactivity indices for the K and L regions of different hydrocarbons are given so that these may be compared with their observed carcinogenic activity

    (a) anthracene
    (b) naphthacene
    (c) 1,2-benzanthracene
    (d) 1,2,5,6-dibenzanthracene
    (e) 3,4-benzpyrene
    (From Pullman[55])

neither of these fulfils the condition for carcinogenic activity. If, however, the additional ring is added not in a linear fashion, but in a lateral position, so that 1,2-benzanthracene is formed as in Fig. 4.24(c), then the K region index falls appreciably to below the critical value of 3·31. The index of the L region now rises, but does not actually reach the threshold value of 5·66, although it is clear that the addition of a new ring in this position has helped to make the molecule of a carcinogenic form, so far as both the K and L region activities are concerned. Experimentally, this particular benzanthracene can be classified as a 'borderline carcinogen'.

If a second benzene ring is added in the lateral position, as for the case of the 1,2,5,6-dibenzanthracene of Fig. 4.24(d), then it is seen that the L region index is now increased to 5·69, and thus noticeably above the critical value, while the K region still remains below 3·31. Both of the conditions for carcinogenesis are, therefore, fulfilled in this molecule, and it is found experimentally to be carcinogenic. It should be realized that the theory refers to the molecular orbital exchange integral for the L region only if such a region actually exists in the molecule, and the case of Fig. 4.24(e) is of particular interest in this respect. In this molecule, a second additional benzene ring has been added to the original anthracene, but in such a way that the L region is completely suppressed. This compound, known as 3,4-benzpyrene, thus fulfils the first condition of having a very active K region with an index noticeably below 3·31 and does not have an L region at all. It should, therefore, be a strong carcinogen and experimentally this is found to be so. These examples can be multiplied and good correlation is nearly always obtained between the predictions of the theory and the experimentally observed carcinogenic activity.

It should be stressed that, in this theory, the indices which are being measured are related to the exchange integrals of the ordinary molecular orbitals of the paired electrons, and do not refer to unpaired spin densities, as discussed in the stable radicals earlier in §4.2. Thus there is no reason why these molecules by themselves should have any unpaired electron density associated with them, and in the absence of such unpaired electrons, they will not be paramagnetic and hence not observable by electron spin resonance techniques. This fact is again supported by most of the experimentally obtained measurements to date,[56] in that there are no consistent electron spin resonance results which indicate that carcinogenic material by itself has noticeably greater unpaired electron concentration associated with it.

On the other hand, one or two other authors have put forward specific proposals suggesting that some forms of carcinogenic

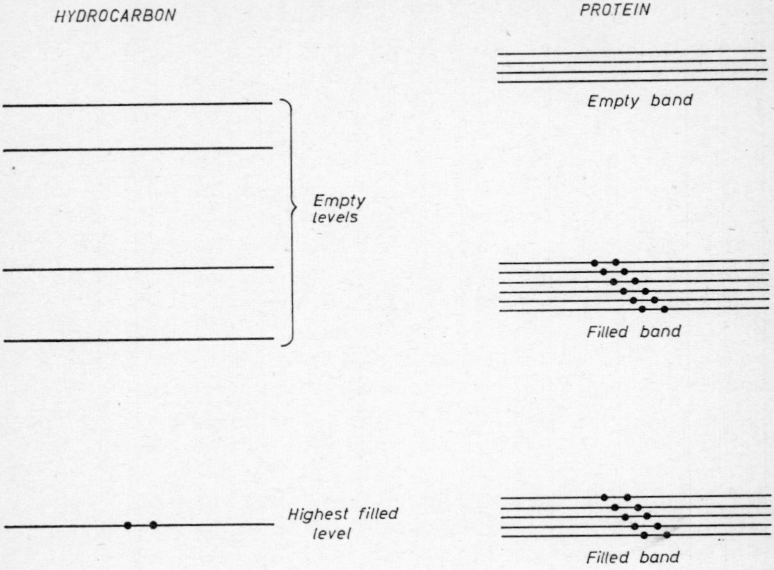

HYDROCARBON

PROTEIN

Empty
levels

Empty band

Filled band

Highest filled
level

Filled band

(a) Before electron transfer

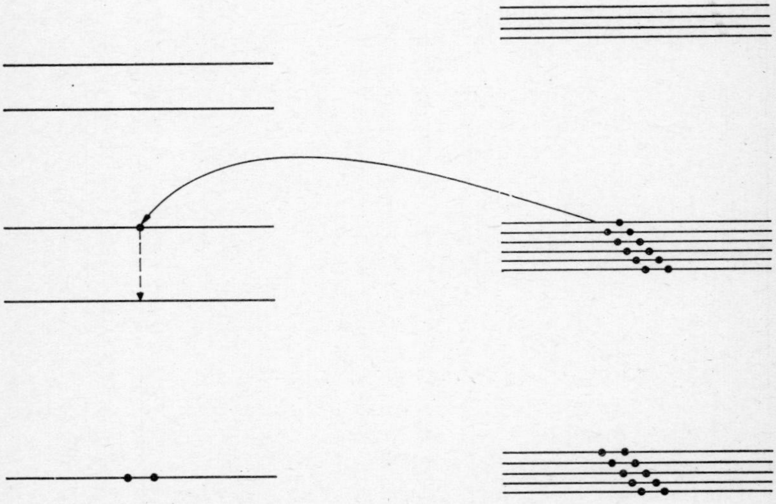

(b) After electron transfer

activity might be correlated with more particular reaction mechanisms, such as electron or excitation transfer between the aromatic molecules and the biochemical cellular receptors. Thus Mason[57] has postulated that the carcinogenic activity depends directly on the existence of electron transfer from the highest filled band of the protein energy-level system to one of the empty levels of the associated hydrocarbon. Thus, in order for such a transfer to occur, an unfilled level of the hydrocarbon must match in energy with the highest filled level of protein itself. Such a theory is illustrated schematically in Fig. 4.25, and it is then assumed that the transfer of an electron from the protein to the hydrocarbon levels will release an unpaired electron in the band structure of the protein itself, and this active agent may then be responsible for the carcinogenic activity.

It is, therefore, evident that the crucial point of this theory is the matching of the two systems of energy levels. The energy gap of the protein band structure is of the order of $1 \cdot 10$ to $1 \cdot 18$ in units of β (the resonance exchange integral of the molecular orbital theory). One simple and direct test of this theory is, therefore, to see if there is any direct correlation between the carcinogenic activity of these hydrocarbon molecules and the width of the gaps in their energy-level systems. Unfortunately, agreement on this particular point has been rather hard to find, since Mason himself has claimed that the results do support his postulates quite significantly,[57] whereas others, such as the Pullmans, have attempted to show that there is in fact no correlation at all.[58]

Another partial test for this theory was also attempted by direct use of electron resonance techniques.[59] In this series of experiments, the proteins themselves were taken in the absence of the hydrocarbon carcinogens, and these were then irradiated with ultra-violet light of various wavelengths in order to try and determine the energy gaps of their energy-level system more directly. These electron resonance measurements were, in fact, carried out not only with a variety of ultra-violet wavelengths, but also under various temperature conditions, and the net result of these experiments was to suggest that quite significant unpaired electron concentrations were, in fact, produced when ultra-violet energy corresponding to the critical gap index was used in the irradiation

---

FIG. 4.25. (*facing*) Carcinogenic activity from matched energy levels
In Mason's theory[57] of carcinogenic activity the unfilled energy level of the hydrocarbon matches the highest filled energy level of the protein as shown in (*a*). Electron transfer then takes place from the protein to the hydrocarbon, as illustrated in (*b*), to release an unpaired electron within the molecular orbital pattern of the protein itself

process. It also became clear, however, that the unpaired electrons that were being detected were not actually in the higher energy level itself, but had become trapped in somewhat lower lying levels, beneath the main band structure. These levels could thus be considered very similar to the impurity levels in a normal inorganic semiconductor, and these electron resonance measurements seem to give fairly striking confirmation of the correctness of the semiconductor band-level theory, as applied to the proteins themselves. Since the crucial point in Mason's theory[57] was, however, the correlation of the hydrocarbon's carcinogenic activity with its particular energy-level structure, these electron resonance measurements on the proteins were unable to give any additional evidence, either in support of, or against, this particular theory of Mason's.

An alternative theory has been put forward by Birks,[60] suggesting that the induction of cancer by conjugated hydrocarbons could be the consequence of the transfer of excitation energy by resonance from tryptophan to the carcinogen, this being followed by the formation of a specific protein–carcinogen complex. This theory thus suggests that the mechanism of carcinogenic initiation is via a dipole–dipole energy transfer rather than the transfer of an electron itself. The crucial parameter in Birks's postulate was thus the value of the overlap integral corresponding to the first excited singlet of the hydrocarbon. Again there has been disagreement on the experimental evidence supporting such a postulate. Birks and his co-workers[60, 61] claimed that there was significant correlation between carcinogenic activity and the magnitude of this particular overlap integral, but Pullman[62] again claimed it was possible to show that no such positive correlation really existed, and that this particular mechanism was, therefore, not borne out by the experimental facts.

The fact that the theories of Mason and Birks have been somewhat discounted does not mean, however, that electron transfer does not play an important part in carcinogenic activity. In fact, Szent-Gyorgi and his collaborators[63] have recently shown that charge transfer complexes are formed by a number of carcinogens and iodine. Moreover, the theoretical results of Pullman can also be interpreted in terms of a charge density distribution in the molecule of the carcinogen, and hence in terms of its capacity to donate electrons at specific points. It should also be noticed that some positive electron resonance measurements have, in fact, been detected in these specific studies, and E.S.R. signals have been reported in a number of charge-transfer complexes, including those with indole, as well as with iodine. In particular, Allison and

TABLE 4.2 Unpaired spin concentrations as measured by E.S.R.
in complexes of carcinogenic and related non-carcinogenic compounds with tetracyanoethylene[64]

Donor molecule	Unpaired spin concentrations of complex spins/g	Unpaired spins per mole complex	Ratio of spin concentration
Azobenzene	7×10^{15}	3.7×10^{-6}	1
1,2-benzpyrene	8×10^{16}	5.2×10^{-5}	11
3,4-benzpyrene	1.8×10^{17}	1.1×10^{-4}	25
1,2-benzanthracene	2.0×10^{17}	1.2×10^{-4}	27
9,10-dimethyl-1,2-benzanthracene	2.4×10^{18}	1.3×10^{-3}	340
4-dimethylaminobenzene	2.5×10^{18}	1.4×10^{-3}	350

Nash[64] have studied 1 : 1 molecular complexes of various carcinogenic and related non-carcinogenic molecules with tetracyanoethylene. These were sealed into glass tubes under nitrogen and studied in a 100 kc/s X-band spectrometer, and the spin concentrations were measured at room temperature. The results they obtained are summarized in Table 4.2, where the compounds are listed in order of their increasing donor strength. These electron resonance measurements make it quite clear that the observed electron spin concentrations in such complexes rise directly with the donor strength of the molecules concerned.

It would probably be fair to say, as a summary of the position at the moment, that there are fairly strong views on both sides supporting and denying the importance of the role of electron donation and acceptance as a crucial step in carcinogenic activity. There is no doubt, however, whatever the final outcome, that experimental measurements obtained by electron resonance techniques will be of very great assistance in establishing more definite and precise facts on which these theories can be based and assessed.

References

1. Sogo, P. B., Nakazaki, M., and Calvin, M., *J. Chem. Phys.*, 1957, **26**, 1343.
2. Hückel, E., *Z. Physik*, 1931, **70**, 204.
3. Johnston, C. S., Visco, R. E., Gutowsky, H. S., and Hartley, A. M., *J. Chem. Phys.*, 1962, **36**, 1580.
4. Weissman, S. I., *J. Chem. Phys.*, 1956, **25**, 890, and McConnell, H. M., *J. Chem. Phys.*, 1956, **24**, 764.
5. Gordy, W., Ard, W. B., and Shields, H., *Proc. Nat. Acad. Sci.*, 1955, **41**, 983-996.
6. Wertz, J. E., and Vivo, J. L., *J. Chem. Phys.*, 1955, **23**, 2193.
7. Gordy, W., and Kurita, Y., *J. Chem. Phys.*, 1960, **34**, 282-288.
8. Kneubuhl, F. K., *J. Chem. Phys.*, 1960, **33**, 1074-1077.
9. Patten, R. A., and Gordy, W., *Radiation Research*, 1964, **22**, 29-44.
10. McConnell, H. M., Heller, C., Cole, T., and Fessenden, R. W., *J. Amer. Chem. Soc.*, 1960, **82**, 76.
11. Heller, C., and McConnell, H. M., *J. Chem. Phys.*, 1960, **32**, 1535.
12. Atherton, N. H., and Whiffen, D. H., *Molecular Physics*, 1960, **3**, 1.
13. Ghosh, D. K., and Whiffen, D. H., *Molecular Physics*, 1959, **2**, 285.
14. Horsfield, A. J., Lin, W. C., and McDowell, C. A., *J. Chem. Phys.*, 1961, **35**, 757.
15. Henriksen, T., Sanner, T., and Pihl, A., *Radiation Research*, 1963, **18**, 147.
16. Drew, R. C., and Gordy, W., *Radiation Research*, 1963, **18**, 552.
17. Patten, R. A., and Gordy, W., *Proc. Nat. Acad. Sci. U.S.*, 1960, **46**, 1137-1144.
18. Pohlit, H., Rajewsky, B., and Redhardt, A., *Free Radicals in Biological Systems*, pp. 367-372 (Academic Press, New York, 1961).
19. Riesz, P., and White, F. H., *Nature*, 1967, **216**.
20. Gordy, W., and Mizagawa, I., *Radiation Research*, 1960, **12**, 211.

21. Blumenfeld, L. A., and Kalmanson, A. E., *Biofizika*, 1957, **2**, 546, and 1958, **3**, 87.

22. Henriksen, T., *Nat. Acad. Sci. National Research Council Report*, No. 43, p. 89 (Washington D.C., 1966).

23. Hunt, J. W., Till, J. E., and Williams, J. F., *Radiation Research*, 1962, **17**, 703–711.

24. Ingalls, R. B., and Wall, L. A., *J. Chem. Phys.*, 1961, **35**, 370–371, and 1964, **41**, 1112, 1120.

25. Cole, T., and Heller, H. C., *J. Chem. Phys.*, 1965, **42**, 1668–1675.

26. Herak, J. N., and Gordy, W., *Proc. Nat. Acad. Sci.*, 1965, **54**, 1287–1292, and 1966, **56**, 1354–1360.

27. Snipes, W., and Schmidt, T., *Radiation Research*, 1966, **29**, 194–202.

28. Jensen, H., and Henriksen, T., *Radiation Research*,

29. Henriksen, T., *J. Chem. Phys.*, 1962, **37**, 2189–2195.

30. Fessenden, R. W., and Schuler, R. H., *J. Chem. Phys.*, 1963, **39**, 2147.

31. Commoner, B., Heise, J. J., and Townsend, J., *Proc. Nat. Acad. Sci.*, 1956, **42**, 710.

32. Sogo, P. B., Pon, N. G., and Calvin, M., *Proc. Nat. Acad. Sci.*, 1957, **43**, 387, and *Radiation Research Suppl.*, 1959, **1**, 511.

33. Sogo, P. B., Jost, M., and Calvin, M., *Radiation Research Suppl.*, 1959, **1**, 511.

34. Van de Waals, J. H., and de Groot, M. S., *Molecular Physics*, 1959, **2**, 333, and 1963, **6**, 545.

35. Ptak, M., and Douzou, P., *Comptes Rendue*, 1963, **257**, 438.

36. Shiga, T., and Piette, L. H., *Photochem. Photobiol.*, 1964, **3**, 223.

37. Shiga, T., Mason, H. S., and Simo, C., *Biochemistry*, 1966, **5**, 1877.

38. Ingram, D. J. E., and Bennett, J. E., *Phil. Mag.*, 1954, **45**, 545, and Ingram, D. J. E., Tapley, J. G., Jackson, R., Bond, R. L., and Murnaghan, A. R., *Nature*, 1954, **174**, 797.

39. Uebersfeld, J., Etienne, A., and Combrisson, J., *Nature*, 1954, **174**, 614.

40. Ingram, D. J. E., *Faraday Society Discussions*, 1955, **19**, 179.

41. Hirsch, P. B., *Proc. Roy. Soc. A.*, 1954, **226**, 143.

42. Castle, J. G., *Phys. Rev.*, 1953, **92**, 1062; 1954, **94**, 1410; and 1954, **95**, 846.

43. Ingram, D. J. E., and Tapley, J. G., *Chemistry and Industry* (London), 1955, 568.

44. Uebersfeld, J., and Erb, E., *J. Physique et Radium*, 1955, **16**, 340.

45. Austen, D. E. G., and Ingram, D. J. E., *Chemistry and Industry* (London), 1956, 981.

46. Singer, L. S., *Proc. Fifth Carbon Conference*, p. 37 (Pergamon Press, Oxford, 1962).

47. Singer, L. S., and Lewis, I. C., *Carbon*, 1964, **2**, 115.

48. Lewis, I. C., and Singer, L. S., *Carbon*, 1967, **5**, 373.

49. Mrozowski, S., *Carbon*, 1965, **3**, 305.

50. Arnold, G. M., *Carbon*, 1967, **5**, 33.

51. Arnold, G. M., and Mrozowski, S., *Carbon*, 1968, **6**, 243.

52. Lyons, M. J., Gibson, J. F., and Ingram, D. J. E., *Nature*, 1958, **181**, 1003.

53. Pullman, A., and Pullman, B., *Advances in Cancer Research*, 1955, **3**, 117, and *Nature*, 1962, **196**, 228.

54. Bhragava, P. M., and Heidelberger, C., *J. Amer. Chem. Soc.*, 1956, **78**, 3761.

55. Pullman, A., and Pullman, B., *Nature*, 1963, **199**, 467, and *Quantum Biochemistry* (Wiley, New York, 1963).

56. Commoner, B., and Ternberg, J. L., *Proc. Nat. Acad. Sci.*, 1961, **47**, 1374, and Nebert, D. W., and Mason, H. S., *Cancer Res.*, 1963, **23**, 833.

57. Mason, R., *Nature*, 1958, **181**, 820, *Brit. J. Cancer*, 1958, **12**, 469, and *Faraday Soc. Disc.*, 1959, **27**, 129.
58. Pullman, A., and Pullman, B., *Nature*, 1962, **196**, 228.
59. Allen, B. T., and Ingram, D. J. E., *Free Radicals in Biological Systems*, p. 215 (Academic Press, New York 1961).
60. Birks, J. B., *Nature*, 1961, **190**, 232.
61. Memory, J. D., *Biochim. Biophys. Acta*, 1962, **64**, 396.
62. Pullman, A., and Berthod, A., *Biochim. Biophys. Acta*, 1963, **66**, 277.
63. Szent-Gyorgi, A., Isenberg, I., and Baird, S. L., *Proc. Nat. Acad. Sci.*, 1960, **46**, 1445.
64. Allinson, A. C., and Nash, T., *Nature*, 1963, **197**, 758.
65. Commoner, B., Heise, J. J., Lippincott, B. B., Norbert, R. E., Passonneau, J. V., and Townsend, J., *Science*, 1957, **126**, 57.
    Commoner, B., *Light and Life* (Johns Hopkins Press, 1961), pp. 356–377.

Chapter 5

Enzyme Studies

THE APPLICATION OF E.S.R. TO THE STUDY OF ENZYMES

The study of enzymes and their various reactions with biochemical and biological systems probably illustrates the various applications of electron spin resonance in the biological field better than most other studies. There are three very important features associated with most enzymatic activity; all can be directly detected and measured by the electron spin resonance technique.

In the first place, there is the production of free radicals, which are normally associated with the enzyme during its active phase, and such free radical signals can of course be readily detected by E.S.R. In the second place, a large number of important enzymes also have metal ions associated with them, such as iron, copper or molybdenum, and the enzymatic activity is often associated with valency changes in these metal ions. Since such valency changes also often produce an unpaired electron configuration around the metal ion itself, they can again be followed and characterized by E.S.R. techniques. The third and probably most important point is that the essential mechanism of enzyme activity, which is basically of a catalyctic nature, is intimately bound up with the kinetics of the chemical changes with which enzymes are associated; hence the study of such kinetics is an extremely important part of any study of enzymes in action. It is in this field at the moment that electron spin resonance is showing itself particularly useful, since even rapid changes in the concentration of both the free radicals and the metal ions can now be followed continuously by the E.S.R. spectrometers.

In fact the great importance of electron resonance techniques in these particular applications is the fact that they can monitor changing free radical concentrations and changing valency states of the metal ions at the same time, hence they not only show which particular valencies are directly associated with the activity of the enzyme, but also in what order the various reactions take place and when secondary reactions start to replace the primary steps. Quite a large number of different techniques are therefore brought into operation in such enzyme studies, since not only have the different types of spectra associated with free radicals and metal ions to be studied, but also a variety of experimental methods need to be applied to determine the kinetics of the various stages. Thus, in some studies on enzymes the continuous-flow methods described in Chapter 3 have been employed, whilst in others the sudden freezing technique has been found more suitable. The way in which these different techniques have been used is illustrated by the series of different examples given in later sections of the chapter on the various enzyme systems that have been studied by E.S.R.

The kinds of electron resonance spectra that are observed from free radicals in biochemical and biological systems, and their interpretation, have already been discussed in some detail in the last chapter, and therefore no further introductory comments on these will be necessary. On the other hand, no detailed consideration has yet been given to the spectra that are observed from the unpaired electron associated with the metal atoms or ions. A brief general consideration of the spectra to be expected from such paramagnetic atoms is therefore given, before the signals obtained from such metal ions in enzymes are considered in detail.

§5.2

E.S.R. SPECTRA FROM METAL IONS

5.2.1 *g-values of metal ions*

It has already been pointed out that two of the most characteristic parameters associated with any electron resonance spectrum are first its g-value and second any hyperfine splittings that are observed. In the case of free radical spectra, the g-values are nearly always very close to the free spin value 2·0023, since the unpaired electrons are moving in highly delocalized molecular orbitals and are therefore not interacting significantly with the orbital motion of the electrons associated with any one atom. In the introductory sections, it was pointed out that the divergence of the g-value from 2·0 is normally a measure of the amount of additional magnetic moment contributed by the orbital motion of the electron, as compared with its spin contribution. Although the detailed expressions for the g-values depend on the particular energy levels of the paramagnetic atom concerned, their general form can be written as

$$g = 2 \cdot 0023 \left[1 - f\left(\frac{\lambda}{\Delta}\right) \right]$$

or
$$g = 2 \cdot 0023 \left[1 - \alpha^2 . f \left(\frac{\lambda}{\Delta} \right) \right] \tag{5.1}$$

if covalent bonding is present.

In this equation, λ is the parameter measuring the strength of the spin orbit coupling of the particular paramagnetic atom being considered, while Δ is the energy level splitting between the ground state in which the electron resonance is being observed and the next appropriate higher orbital state of the atom in the particular configuration in the molecule under study. The term $f(\lambda/\Delta)$ indicates that the divergence of the g-value from $2 \cdot 0023$ is expressed as some function of the ratio of these two parameters; quite often, this function is just of a simple linear form. General details of the theory of these g-values, and the way in which this particular function can be derived for the different cases, is considered in detail in other books on the subject;[1] since this theory is somewhat complex it will not be considered in detail here. In the biochemical and biological applications, these g-values are used much more as a means of characterizing a particular metal atom and its valency state than for determining the detailed energy level system of the atom concerned. Hence a knowledge of the form of the function $f(\lambda/\Delta)$, and the particular values it takes for the different atoms, is the main concern of workers in this field. As an example, the case of copper can be quoted, where the $f(\lambda/\Delta)$ function takes the simple form of

$$f \left(\frac{\lambda}{\Delta} \right) = \frac{4\lambda}{\Delta} \tag{5.2}$$

for the cupric atoms when they are ionically bound to the surrounding ligands;[2] this value is reduced by a covalency factor, α^2, if significant covalent bonding is also present.

The actual expression for this function depends, in fact, not only on the valency state, and hence the number of unpaired electrons associated with the metal ion in question, but also on the nature and symmetry of the crystalline, or molecular, electric field in which the ion is located. This dependence arises from the effect of these internal electric fields on the orbital energy level splittings of the electrons, and hence the molecular fields determine not only the value of Δ, but also the quantum characterization of the ground level itself. Thus, in case of a Cu^{++} ion in a position having octahedral symmetry with some tetragonal distortion (as is often the case in simple inorganic salts), the ground state level is then $d_{x^2-y^2}$ and the Δ of $f(\lambda/\Delta)$ is the splitting between this ground state and the d_{xy} level. If, however, the symmetry of the surroundings of the copper ion were tetrahedral, rather than octohedral, then the orbital energy level pattern would have been inverted, with the d_{xy} level lowest, and with a different magnitude for Δ, and expression for $f(\lambda/\Delta)$.

It is therefore evident that not only can the valency state of the metal ion be determined from its g-value, but also the symmetry of its molecular surroundings.

It would appear from the foregoing statements that the analysis and characterization of E.S.R. spectra for metal ions would be extremely simple, since each different valency state of each different metal ion would have its own characteristic g-value, and hence these could all be readily identified immediately from the resonance field position of the observed absorption line. Unfortunately, however, the position is not quite so straightforward as this, since there is often a considerable amount of anisotropy associated with the g-values from such metal ions. This anisotropy arises from the competition between the effect of the externally applied magnetic field on the magnetic moments of the electrons, and the effect of the internal molecular or crystalline electric field also acting on the electrons via the spin-orbit coupling mechanism. As a result of this, the divergence of the g-value from 2·0 will vary with the angle between the directions of these two fields.

In other words the expression for the function $f(\lambda/\Delta)$ in equation (5.1) is also dependent upon angle; again, as a specific example for the case of Cu^{++}, the g-value variation can be represented by a $g_{\parallel}$ of 2·15, and a $g_{\perp}$ of 2·40. In this case we are assuming that the crystalline, or molecular, electric field has axial symmetry, and the $g_{\parallel}$ represents the g-value observed when the applied external magnetic field is pointing along the direction of this axial symmetry, whereas $g_{\perp}$ represents g-value observed when it is in any direction at right angles to this axis. It may also be noted that the spin orbit coupling parameter, λ, for Cu^{++} is equal to $-850\ cm^{-1}$, and the negative sign explains the shift of the g to a value above 2·0. Intermediate g-values at an angle θ to the axis of symmetry will then be given by the expression

$$g_\theta{}^2 = g_{\parallel}{}^2 \cos^2 \theta + g_{\perp}{}^2 \sin^2 \theta \qquad (5.3)$$

If single crystals are being studied, the variation of such g-values with angle can be followed directly as the externally applied magnetic field changes in orientation with respect to the crystalline axis. However, in the vast majority of cases in which biochemical or biological systems are concerned, single crystals are not available, and, if kinetic measurements are to be made, they must be made on the reacting systems in solution, as close to *in vivo* conditions as possible. The question therefore arises of the kind of electron resonance spectra that will be obtained if such a spread of g-values exists, corresponding to all the different orientations of the different molecules within the liquid or polycrystalline specimen. The net effect of such a spread of g-values is illustrated in Fig. 5.1. In Fig. 5.1(a) the actual absorption curves, spread across

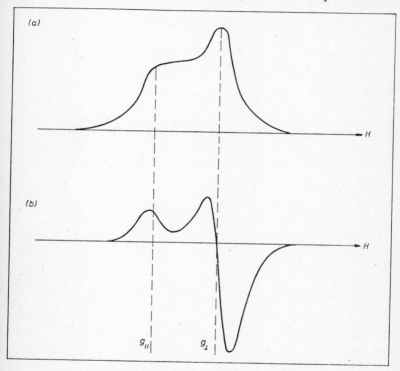

FIG. 5.1. E.S.R. absorption curves for spread of g-values with axial symmetry
(a) Absorption curve obtained from solution of typical Cu^{++} salt
(b) First-derivative recording of the same solution as (a)

the different angles, are indicated for a typical copper spectrum, and in Fig. 5.1(b) the resultant first derivative recording is also plotted since this is the one which is normally observed on the electron resonance output. It will be seen from this figure, which is drawn for the case of an axially symmetric internal molecular field, that there is in fact very much more absorption concentrated near the $g_\perp$ direction than near the $g_\parallel$ direction. This can be readily appreciated when it is realized that, in a three-dimensional variation, there will be many more distributions pointing close to the whole plane of directions represented by the perpendicular g-value than pointing close to the one direction represented by the $g_\parallel$ axis. It is therefore possible to differentiate quite clearly between these two extreme g-values, and the magnitudes can be read off quite specifically from such recorded spectra, even when liquid or polycrystalline material is being studied.

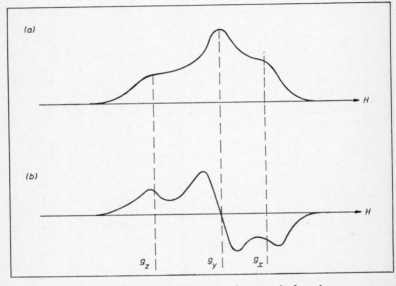

FIG. 5.2. E.S.R. absorption curves for spread of g-values without axial symmetry

(a) Absorption curve from Cu^{++} in an orthorhombic site
(b) First-derivative recording of the same spectrum, showing how principal g-values can still be identified

The situation is slightly more complex when the site of the metal ion in the molecule does not possess axial symmetry, but has a lower order of symmetry, such as orthorhombic, or others in which there is no equivalence between all the directions in the perpendicular plane. In this general case the g-value variation can be represented by three principal values normally denoted by g_x, g_y and g_z; g_z being that along the parallel axis and therefore equivalent to the $g_\parallel$ in the previous case, and g_x and g_y being the maximum and the minimum values in the plane perpendicular to this axis.

The observed absorption lines and first derivative recordings which are produced in this case are shown in Fig. 5.2, and are represented in the same way as in the previous figure. It will be seen that these curves now have three definite turning points, and so, again, the three different *g*-values can be measured from them, even though solutions and not crystals are being studied.

In the above considerations it has been assumed that the *g*-value in solution was identical with that in the crystalline solid state, but smeared out by the random orientation of the molecules in the liquid. This is indeed generally the case, but it is also possible, if the molecular tumbling motion in the solution is high, for an averaging effect to take place, as discussed earlier for the anisotropic hyperfine splitting which is averaged to zero in most solvents. The averaging of the *g*-value divergence from 2·0 is not quite so readily effected in the liquid state however, since the energy splitting, as measured in frequency units, is much larger than that normally associated with hyperfine splittings, and for the larger molecules met with in biological studies, the tumbling motion is often not rapid enough for this averaging to take place. Such a possibility is an important effect to remember however, and observed *g*-value variations should always be compared with solvents, or solutions, of approximately the same viscosity, when allocations to uncertain atoms are being made.

It should also be pointed out that in certain cases, and particularly those in which the metal ion is in a spectroscopic singlet state, which indicates that there is no orbital motion associated with the ground state of this atom, the *g*-value will not shift from the free spin value and hence an isotropic *g* of about 2·0 is observed. One such case is that of the manganese Mn^{++} ion, which is often present as an impurity in buffer solutions. This has an isotropic *g*-value close to that of the free spin, and well-resolved hyperfine lines, consisting of a very characteristic six-line pattern, are normally observed spread on either side of the free spin *g*-value. In other cases, however, as described in the examples which follow in the later sections, quite significant shifts from the free spin value are observed.

5.2.2 *Hyperfine structure in metal ions*

The general features of the hyperfine patterns observed in electron resonance have been discussed in the introductory chapters, and in particular in §1.5. The hyperfine patterns observed from free radical specimens have also been considered at greater length in the last chapter, but it was pointed out in the introductory section that there is a very significant difference between the hyperfine pattern normally observed from a metal ion, as compared with that observed from a free radical. In the latter case, the electron is normally moving in a delocalized orbital and hence is interacting with several nuclei at once. In the case of most spectra observed from metal ions, however, the unpaired electron is usually very considerably localized on the metal ion itself, and therefore the main hyperfine pattern arises from any interaction with the nuclear spin of that particular ion.

Thus, in a simple case of say an inorganic salt containing copper ions, there would be a direct interaction of the unpaired electron with the nuclear spin $I = 3/2$ of the copper atom, and hence a $(2I + 1)$, i.e. a four-line hyperfine pattern, is observed. This hyperfine structure has already been illustrated in Fig. 1.11(a), which shows the absorption lines actually obtained from a single crystal containing a Cu^{++} ion, and in Fig. 1.12(b), the spectrum observed from an enzyme system containing copper ions in solution was given as a further illustration of such hyperfine splittings. The additional feature, immediately apparent from Fig. 1.12(b), is that the four-line hyperfine pattern associated with the $g_{\parallel}$ position can still be readily identified, even from the random orientation of the metal ions in solution. This retention of the characteristic hyperfine pattern by a metal ion in solution, is a very considerable help in biochemical and biological studies, since characterization of the metal ion taking place in the enzyme activity can then be made directly, under conditions which are very similar to those existing *in vivo*.

It is possible, however, even if the electron is mainly localized on the single metal ion, for its molecular orbital to move out and embrace some of the surrounding ligand atoms, and in this case there may well be additional hyperfine interaction from the nuclear spins of these ligand atoms which surround the metal ion itself. One common case that often arises in biochemical and biological studies is that of a square of four nitrogen atoms surrounding the metal ion. The porphyrin and haem planes form very good examples of such a system, since both of these contain a central metal atom with four nitrogens surrounding it, and this forms part of a much larger planar molecular unit. A very similar molecule is that of the phthalocyanines, and their structure has already been quoted as an example and is illustrated in Fig. 1.14(a). It is possible for quite a large number of different metal ions to be substituted into the central position of this molecule, and, if the specific case of copper is again taken, the hyperfine pattern that is then observed is shown in Fig. 1.14(b). This hyperfine splitting was obtained from electron resonance studies on single crystals, and hence all the angular information is present, and no averaging or broadening effects have taken place. A brief discussion of this spectrum, in §1.5, indicated that the extra splitting now observable on each of the four copper ions, was due to the superhyperfine structure arising from the interaction of the unpaired electron with the four nitrogens around the copper ion.[3] Moreover, the magnitude of the splitting of the superhyperfine structure, compared with that of the splitting between the copper hyperfine components, will give a direct measurement of the proportion of the unpaired electron spin

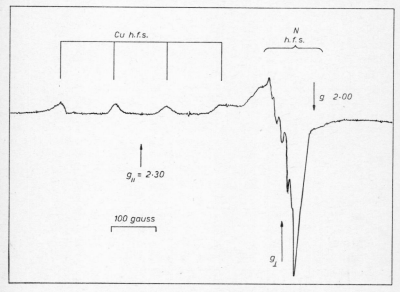

FIG. 5.3. E.S.R. from solution of copper conalbumin
The four hyperfine components centred on the $g_{\parallel}$ position are from the copper nucleus, but the finer splittings around the $g_\perp$ position are the superhyperfine lines from the nitrogen atoms surrounding the copper (From Windle *et al.*[4])

density present at these ligand atoms, and thus of the way in which the molecular orbital is distributed over the molecule itself.

Even these much finer superhyperfine components can still be retained and resolved in studies on solutions, and a good example of this is shown in Fig. 5.3, where the spectrum observed from aqueous solution of copper conalbumin is shown.[4] The four individual lines around the $g_{\parallel}$ position, arising from the copper atoms themselves, are shown clearly at the left of the figure, while on the right-hand side, the large single line represents the higher

TABLE 5.1 Hyperfine patterns and typical g-values of metal ions in enzymes

Metal atom	Isotope with abundance	Nuclear spin	Hyperfine pattern	Typical g-values
Copper	Cu^{63}(70%) Cu^{65}(30%)	$\frac{3}{2}$ $\frac{3}{2}$	Both give four lines which very nearly overlap, to give appearance of one single set.	$g_\perp = 2.3$ $g_\parallel = 2.05$ for Cu^{++}
Iron	Fe^{57}(2%) Even isotopes (98%)	$\frac{1}{2}$ —	Very weak hyperfine pattern compared with single line—not readily observable.	Low spin Fe^{+++} has g-values close to 2.0 (see next chapter for high spin cases)
Manganese	Mn^{55}(100%)	$\frac{5}{2}$	Six-line isotropic hyperfine pattern well resolved in most solids, liquids and glasses.	$g = 2.00$ for Mn^{++}
Molybdenum	Mo^{95}(16%) Mo^{97}(9%) Even isotopes (75%)	$\frac{5}{2}$ $\frac{5}{2}$	Six-line hyperfine pattern observed about main central line due to even isotopes. Hyperfine pattern can be enhanced by use of enriched Mo^{95}.	$g_\parallel = 2.02$ $g_\perp = 1.95$ for Mo^{5+}
Cobalt	Co^{59}(100%)	$\frac{7}{2}$	Eight-line hyperfine pattern observed but often highly anisotropic.	g-values vary widely with valency state and particular location of cobalt atoms
Vanadium	V^{51}(99.7%)	$\frac{7}{2}$	Eight-line hyperfine pattern often isotropic and well resolved.	$g = 2.00$ for V^{++}

intensity associated with the $g_\perp$ components. It is seen however that this line has quite a number of much finer hyperfine components superimposed on it, and this superhyperfine splitting associated with the peak at the $g_\perp$ is due to an interaction of unpaired electron with the nitrogen surrounding the copper atoms, since the copper hyperfine structure itself has collapsed to a very small value at this orientation. This example again shows that quite often a large amount of quantitative data on hyperfine splittings can still be obtained, even from aqueous solutions of the reacting enzymes.

The nuclear spins of the isotopes of the metal atoms of interest in enzyme studies are listed in Table 5.1, so that this information, together with a summary of the resultant hyperfine patterns and also typical g-values, might be readily available for comparison purposes. The ways in which these hyperfine patterns can be used to identify and follow the different metal ions and their valency states in enzyme reactions, is probably best illustrated by the specific examples which follow in the succeeding sections of this chapter.

§5.3

GENERAL FEATURES OF ENZYME REACTION

Before considering specific examples of the way in which electron resonance has been applied to the study of different enzyme systems, one or two general points about such reactions will first be considered. Enzymes have some general features which are very similar to other systems that have been studied by electron resonance, whereas in other aspects they are unique and possess very specific properties.

Thus the first essential feature of an enzyme is that it acts as a catalyst for a chemical or, in this case, a biological reaction, and, as with all catalysts, its essential feature is to increase the rate of the chemical reaction. In such reactions the concentration of a catalyst itself remains unchanged in the ideal case, but it is often found in such catalytic activity that the rate of the catalysed reaction is directly proportional to the concentration of the catalysts present, and this is, in fact, nearly always the case with enzymes. In fact most enzyme-catalysed reactions do not occur at any appreciable rate in the complete absence of the enzyme, so that a plot of the rate of the action against the enzyme concentration is then simply a straight line passing through the origin.

The enzymes are remarkable, however, not only for their very high effectiveness as catalysts, but also for the fact that they exhibit a very definite specificity with respect to the particular compound which is to be catalysed. These substances are called the substrates, and each enzyme will normally act as a catalyst only for its own specific substrate. In fact, however, quite a wide range of such specificity of reaction does occur, and enzymes can be roughly grouped into four different headings in this respect. In the first group there are those with what is termed *absolute specificity*, when the enzyme will bring about a reaction in only one single substrate. An example of this is the case of urease, which will catalyse the hydrolysis of urea to ammonia and carbon dioxide, but is quite ineffective on other substances, even those which have closely related chemical structure.

In the second group of enzymes are those with what is called a *group specificity*. These will act on a series of substrates, but the substrates must be similar and possess specific types of groupings if the enzyme is to act on them. Thus, the enzyme pepsin will hydrolyse certain peptide linkages, but only if aromatic groups are present in certain positions with respect to the peptide linkage. In the third group of enzymes, the specificity is associated with a certain type of reaction or chemical linkage; and, in the fourth group, the specificity is associated with stereo-chemical form, rather than chemical reactions. Thus, for example, lactic dehydrogenase will catalyse the oxidation *l*-lactic acid, but it does not catalyse the oxidation of the *d*-form.

As well as being classified by their specificity, enzymes can also be classified and characterized with respect to the types of reactions which they catalyse. Such a classification can be divided into about six main groups, which are briefly summarized below.

1. The *hydrolytic enzymes* which catalyse reactions of the type indicated by

$$AB + H_2O \rightarrow AOH + HB$$

This group itself can be subdivided into four further classes, according to the particular chemical group or linkage which is being hydrolysed in this way.

2. The second group contains the *phosphorylytic enzymes* or the *phosphorylases*. This group catalyses reactions in which a molecule of phosphoric acid replaces the molecule of water in the last equation, and this molecule minus one proton is added to the substrate as shown:

$$AB + HO{-}\underset{\underset{OH}{|}}{\overset{\overset{O}{|}}{P}}{-}OH \rightarrow AO{-}\underset{\underset{OH}{|}}{\overset{\overset{O}{|}}{P}}{-}OH + HB$$

3. The third group of enzymes are called the *oxidative enzymes* and are those concerned with oxidative processes of various types. They can again be subdivided into different classes, which include the *dehydrogenases*, which catalyse the removal of two hydrogen atoms from a substrate molecule. The second subgroup contains the *oxidases*, which catalyse an oxidation process in which hydrogen atoms are transferred directly to molecular oxygen, and the third subgroup are known as the *oxidative deaminases*, which act specifically upon amino compounds, bringing about oxidation together with the elimination of a molecule of ammonia.

The fourth and fifth main groups are then termed *the adding enzymes* and the *transferring enzymes*, respectively, in that they catalyse simple addition reactions on the one hand, and reactions in which groups are interchanged on the other. Finally the sixth group can be classified as the *isomerizing enzymes*, which catalyse various types of isomerization process, and a large number of these involve phosphorylated substances.

It will be appreciated that this particular type of classification is only for convenience, and not of any fundamental significance, but it does serve to illustrate the different kinds of chemical and biochemical reaction, in which these specific types of catalyst can take part.

In the earlier studies of enzyme reactions, attention was normally focused on methods of measuring the kinetics of the enzyme reactions, and then the results of such kinetic studies were used to deduce the detailed reaction mechanism and the kind of intermediate products that were being formed in the process. Thus, in some of the early studies, the rates of enzyme-catalysed reactions were measured for different concentrations of the substrate and various types of behaviour were observed. By far the most common was one in which the rate of reaction increased linearly with substrate concentration for low values of this concentration, but at high substrate concentration, the rate became independent of this, and thus the kinetics changed from the first order to a zero order. This type of behaviour was first interpreted theoretically by Michaelis and Menten,[5] who proposed that the enzyme reaction was taking place in two well-defined steps, the first of which was the formation of a complex between the enzyme and the substrate, while the second corresponded to the decomposition of this complex into the products of the reaction. The change from the first-order kinetics to the saturation condition can then be attributed to the fact that at low substrate concentrations most of the enzyme molecules are, in fact, present as such, and only a fraction are combined with the substrate molecules. When the substrate concentration rises to a high value, however, practically all the enzymes will be in the form of a complex with the substrate, and further increase in the substrate concentration cannot, therefore, increase the concentration of the complex itself any further, and hence the rate of reaction is no longer affected. Detailed studies of most enzyme systems show that the situation is somewhat more complex than this simple picture would suggest, but this kind of basic argument shows how important kinetic studies are in the investigation of enzyme reactions.

The great limitation of the previous chemical techniques was that they were generally unable to measure the concentration of any of the intermediate complexes at all accurately, and the kinetic studies were based on measured concentrations of the initial and end products. The intermediate steps were then deduced from models which would fit the observed kinetic pattern as well as possible. The great advantage of electron spin resonance, in this connection, is that the intermediate steps are normally those with which the unpaired electrons are associated, and hence these can be measured and characterized specifically by the electron resonance technique. Thus much more direct information is available on the actual steps in the enzyme catalytic process, and the kinetics of primary and secondary reactions can be distinguished quite straightforwardly. The way in which this is carried out can probably be illustrated best by considering specific examples. The remaining sections of this chapter will therefore be concerned with the way in which E.S.R. has been employed to investigate the mechanisms of specific enzyme reactions.

§5·4

STUDIES ON XANTHINE OXIDASE

It has already been seen that one of the main groups of enzymes are known as the *oxidative enzymes*, and this includes both the dehydrogenases and the oxidases within it. These two groups between them do, in fact, contain some of the most important and interesting enzymes that exist and also, not surprisingly, quite a large number, therefore, of those that have been studied in detail by electron resonance techniques. One of these is the xanthine oxidase of milk, and although this is not necessarily the most important of these oxidases, it has, been studied extensively by electron resonance techniques. A brief summary of the results obtained will, therefore, illustrate the different types of information that can be deduced in such studies.

Xanthine oxidase contains flavin, iron and molybdenum and it was, therefore, expected that there would be both free radical contributions to the intermediates of the enzyme reaction, together with possible valency changes in both the iron and molybdenum. Hence considerable information might be deduced from electron resonance studies of its reaction. It is also an enzyme which fulfils several other useful requirements for electron resonance studies, since it can be prepared from an easily available source. It can also be obtained in high purity and in quite large quantities within a few days, and conditions for stabilizing the enzyme and storing it for several months are known. The first electron resonance experiments on this enzyme were carried out by Bray *et al.*,[6] who were able to show in 1959 that, on reduction, two E.S.R. components could be distinguished. One of these had a g-value very close to that of the free spin, and was, therefore attributed to the flavin radical; while the other had a g-value noticeably shifted from this at 1·97 and attributed to Mo^{5+} (henceforth denoted as Mo(V), signifying molybdenum atoms in the fifth valency state). Further studies[7, 8] on the enzyme then showed that four different electron resonance signals could be distinguished, including the first two attributed to the flavin and Mo(V), and two others which had g-values corresponding to 4·2 and 1·9. Both of these were assigned to the iron atoms, the resonance at 4·2 being that from the oxidized ferric iron, and that at 1·9 being assumed due to the reduced iron. It was then found, however, that the behaviour of the signal at 4·2 was very erratic, and could not be easily correlated with the kinetics of the other components and it was, therefore, assumed that this was due to a contaminant, which seemed highly likely since a large number of biological materials are often contaminated with ferric iron. Further collaborative studies between Bray and Beinert[9, 10] and Palmer were then able to follow the kinetics of the $g = 1·9$ iron signal in much more detail, and in these studies the rapid freezing technique was employed so that the kinetics of the signals could be followed in the millisecond range intervals. In these rapid kinetic studies, at least four different kinds of signals were distinguished, which all showed rates of appearance and disappearance in the millisecond range, suggesting that they were taking part in the enzyme reaction.

A typical plot of the kinetics of these four signals has already been given in Fig. 3.16, as an illustration of the sudden freezing technique in action. Reference to this figure will show that the free radical signal from the flavin rises very quickly during the first 20 milliseconds, turns over sharply, and then decays over the next several hundred milliseconds. The signal labelled D has a turnover point within the first 20 millisecond period, whereas that labelled B does not reach its maximum until about the 50 millisecond mark. Also, the one attributed to the iron atom rises noticeably more slowly and its maximum does not occur until about 100 milliseconds. The actual form of these signals, traced out as first-derivative recordings, is shown in Fig. 5.4, which is taken from a paper by Palmer *et al.*[10] The first signal, labelled A, is that observed from the xanthine oxidase before it reacts with the xanthine itself and only a weak absorption at the free spin value of $g = 2 \cdot 00$ is obtained. The signals labelled from B to E are those obtained at various intervals after the reaction has been initiated, B occurring at 26 milliseconds and E at 1400 milliseconds. The complexity of the trace for the B and C curves shows the overlap of all four electron resonance absorptions, whereas it is seen that some of these die away significantly by the 1400 millisecond time interval, when the signal due to the iron predominates.

In the g-region of $1 \cdot 95$ to $1 \cdot 97$, there is a complicated structure which changes shape as reduction proceeds. At least two components could be distinguished in this complex signal, and these follow different kinetics. In Fig. 5.4 they are designated as α and β on the one hand, and γ and δ on the other. It was then found that the two components could be obtained separately by mixing the enzyme with xanthine at a pH of $6 \cdot 0$, which gave only the α, β components, while a pH of $9 \cdot 6$ gave the γ, δ component. A detailed analysis of the structure of these separate signals suggested that both were due to different chemical species of Mo(V), since the hyperfine structure due to the molybdenum nucleus could be resolved on both of them, and was particularly noticeable for the γ, δ signals.[10] Reference to Table 5.1 shows that the isotopes of molybdenum are about 75 per cent abundant in those with zero nuclear spin, and which therefore give rise to a single unsplit absorption line. There is also a 25 per cent abundance of Mo^{95} and Mo^{97}, both of which possess a nuclear spin of $\frac{5}{2}$, however, and hence produce a six-line hyperfine pattern.

In order to confirm the interpretation of these signals, Bray and Meriwether[11] proceeded to prepare xanthine oxidase which contained an enhanced proportion of the Mo^{95} isotope. Increasing the proportion of this isotope will produce a higher intensity of the signal with the six-line hyperfine pattern, and confirm whether these particular signals are indeed due to the molybdenum, or not. Initial attempts to introduce the artificially enriched molybdenum by chemical exchange methods were not successful, and a direct biological synthesis of the labelled enzyme was therefore then undertaken. The Mo^{95} was incorporated into the xanthine oxidase by injecting a cow with the isotope and then isolating the enzyme

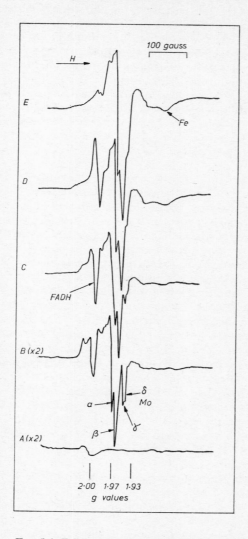

FIG. 5.4. E.S.R. spectra from xanthine oxidase

A. from the xanthine oxidase before the reaction
B-E. at successive time intervals during the reaction with xanthine

These spectra should be compared with the plots in FIG. 3.16, which show how each component of the spectrum varies with time (From Palmer *et al.*[10])

directly from the milk. In order to determine the extent of incorporation of this isotope into the xanthine oxidase, a small amount of radioactive Mo^{99} was also added to the injected solution. It was found that, when a cow weighing 546 kilograms was given a single intravenous injection of 184 milligrams of Mo^{95} as sodium molybdate plus $310\mu C$ Mo^{99}, xanthine oxidase could be isolated from the highly radioactive milkings obtained during the 72-hour period following the injection. Quantitative analysis of the radioactivity showed that 68 per cent of the injected isotopes had been incorporated into the enzyme, and thus about 75 per cent of the molybdenum isotopes in these enzymes would now possess isotopes with nuclear spin.

The experiments on the α, β and γ, δ signals were therefore repeated with these particular enzymes, and the γ, δ signal was first obtained by rapid freezing of solutions of pH 10 and at a time interval corresponding to the very short reaction time of 10 milliseconds. Fig. 5.5 shows the electron resonance signal that was then obtained, the lower trace being that produced by the normal enzyme, whereas the upper trace corresponds to that observed from the enzyme containing enriched Mo^{95}. It is quite evident from this figure, that the γ, δ signal does now possess an intense and well-resolved hyperfine structure of six lines. There is, in fact, considerable anisotropy in the g-value, the g_z, as represented on the trace, corresponds to $2 \cdot 025$, while the g_x and g_y correspond to $1 \cdot 951$ and $1 \cdot 956$ respectively. The six-line hyperfine pattern is clearly seen however, centred around the g_z field position, and the growth in the hyperfine components with the corresponding reduction in the central line, is clearly evident from a comparison of the two curves. These results not only give definite confirmation of the fact that this particular signal is to be attributed to the molybdenum ion, but also, from the actual magnitudes of the g-values, suggest that the Mo(V) is in an octahedral complex with a slight rhombic distortion. It also seems likely that one or more of the ligand atoms may well be sulphur atoms, possibly arising from the cysteine residue in the protein.

The α, β signal was also studied in the same way and the E.S.R. signals obtained from the artificially enriched Mo^{95} enzyme compared with those obtained from the normal enzymes. It was again possible to show that this signal was also associated with the molybdenum ions and obtain information on the way in which these molybdenum atoms might be bound to the surrounding ligands. These spectra are a very good example of the way in which hyperfine structure analysis can be used in such enzyme studies, even when the specimens are in the liquid form, or, as in this case, in a

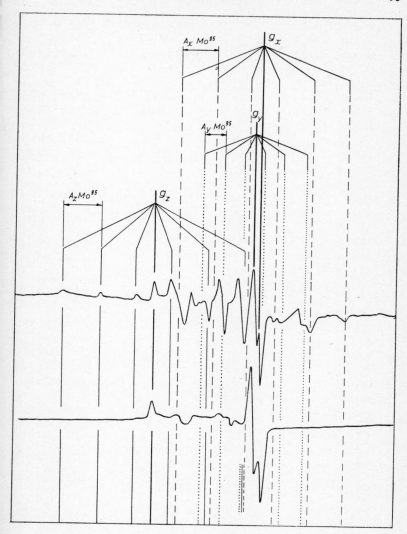

FIG. 5.5. E.S.R. spectra from xanthine oxidase enriched with Mo^{95}
These spectra are for the $\gamma\delta$ molybdenum signal observed at high pH.
The lower trace is from ordinary xanthine oxidase while the upper
trace is from enzymes enriched in Mo^{95}. The increase in intensity of
the hyperfine lines from Mo^{95} can be clearly seen (From Bray and
Meriwether[11])

deep frozen state. The work also illustrates very strikingly the technique of biological enrichment with isotopes possessing nuclear spins, when the normal atoms do not possess these.

The way in which the electron resonance measurements have helped to elucidate the detailed mechanism of the enzymatic activity of xanthine oxidase can be summarized in Fig. 5.6, which is taken from a paper by Bray and his collaborators.[9] In this figure, the different pathways by which electron transfer can take place are illustrated. The various oxidation-reduction systems are aligned

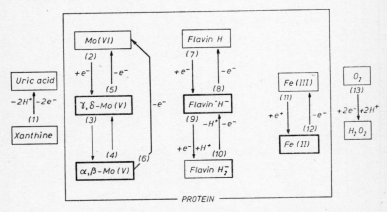

FIG. 5.6. Diagrammatic representation of electron transfer mechanisms in the xanthine oxidase reaction

The different oxidation-reduction systems are aligned vertically, and the various horizontal positions correspond to different levels of oxidation, the top line giving the final oxidized state. Species that can be detected by E.S.R. are drawn in heavy rectangles (From Bray *et al.*[9])

in vertical columns, while the different species within each system are placed on different horizontal levels, and the final oxidized state of the system corresponds to the top line. The actual uptake or loss of electrons or protons is then indicated by heavy arrows labelled with the corresponding electron or proton sign, while the lighter arrows indicate no electron transfer, but a conversion within the same oxidation state. All the intermediates that can be detected by electron resonance are drawn within heavy rectangles, and it is clear from this that a large proportion of the mechanism of the enzyme can thus be followed directly by this technique.

It is probably fair to say that the complete mechanism of the enzymatic activity of the xanthine oxidase still needs to be elucidated, but these particular results serve to show the power of the electron resonance technique, not only in correlating the kinetics of the different components of the enzyme activity, but also identifying them quite precisely from the hyperfine patterns which are produced.

§5·5

STUDIES ON ALDEHYDE OXIDASE

The investigation of xanthine oxidase has been considered as the first example, since it is probably the enzyme that has been studied most by electron resonance, and it does illustrate the various ways in which this technique can be applied. A number of other oxidase enzymes have been studied in a similar way however, as, for example, aldehyde oxidase. This is closely related to xanthine oxidase, both in the composition of its electron transfer mechanism and in its function. Thus electron resonance signals are obtained[12] both from free radicals, which are again presumably from the flavin component, and from the metal ions molybdenum and iron. Moreover a study of the kinetics of these different signals shows that they are somewhat similar to those of xanthine oxidase. More detailed study on these electron resonance signals[13] has shown, however, that there are, in fact, considerable differences between the aldehyde oxidase and the xanthine oxidase, and these differences are mainly concerned with the state of the molybdenum atoms.

Thus the samples of the liver aldehyde oxidase show quite a significant signal, typical of the Mo(V), even in the resting enzyme. This signal has g-values corresponding to a g_z of 1·97 and the hyperfine structure from the molybdenum nuclei can be resolved and thus the signal definitely identified as due to molybdenum ions. Moreover, the intensity of these signals obtained from the resting enzyme suggests that 25 per cent of the total molybdenum is present in this state and thus does not take much part in the reduction or reoxidation cycles associated with the enzyme activity. These two different forms of molybdenum have been differentiated quite strikingly by measurements on relaxation times, as are discussed in the last chapter. Another difference between the aldehyde and xanthine oxidase is that, although the α, β type of signal is obtained in both, the equivalent of the γ, δ signals is not obtained in the aldehyde oxidase spectra. Moreover, the aldehyde oxidase showed very significant changes in its behaviour when the kinetics of the reduced enzyme, and the reduced and incubated enzyme, were compared.

These studies employed both rapid flow and rapid freezing techniques and are a good illustration of the way in which these different methods can be used to distinguish between different features of the enzyme activity. Thus, in the first series of measurements[14] the enzyme was first rapidly mixed with substrate from two syringes, as in a normal continuous-flow technique, and then immediately afterwards was mixed with an oxygenated buffer from a third syringe. These results were then compared with a second set obtained from experiments in which the enzyme was filled anaerobically into the syringe together with the substrate (a process which takes about twenty minutes) and then, after this, mixed rapidly with oxygenated buffer. It was found that, in the case of the reduction together with the incubation period, only the flavin and iron components were readily reoxidized and the molybdenum appeared to be inert. On the other hand, reduction immediately followed by reoxidation produces signals in which all three components, due to the flavin, molybdenum and iron, respond rapidly to the addition of oxygen. These experiments not only confirm the lability of the reduced molybdenum, but also suggest that molybdenum is, in fact, the catalyst which is closest to the substrate. The general conclusions from these experiments on the aldehyde oxidase, suggest that, as in the case of xanthine oxidase, the molybdenum, flavin and iron all participate in the electron transfer from the substrate to oxygen and in the order given.

§5.6

STUDIES ON THE DEHYDROGENASES

It was pointed out in §5.3, when the general features of enzyme reactions were considered, that the main group of enzymes concerned with oxidative processes contained not only the oxidases themselves, but also the dehydrogenases, these being enzymes which catalyse the removal of hydrogen atoms from the substrate molecule. Several such dehydrogenases have now been studied in detail by electron resonance techniques, and some of these results will be collected together briefly in this particular section.

One of the most easily characterizable enzymes in this group is dihydroorotic dehydrogenase, since this contains only iron and flavin in a 2 : 2 stoichiometry per molecule of enzyme, and has a relatively low molecular weight of 60 000. It can moreover be obtained in the crystalline form, and hence readily purified. It also has a distinct optical spectrum, and as a result combined measurements of both the optical and E.S.R. spectra have been made by Aleman *et al.*[15] Rapid freezing techniques were employed using low temperature electron resonance and reflectance spectroscopy at the same time, and a direct comparison of these two

methods of study is shown in Fig. 5.7, which summarizes a set of
kinetic studies on the enzyme solution after rapid mixing with an
equal volume of millimolar DPNH. (i.e. Reduced Diphospho-
pyridine Nucleotide). The percentage reduction is plotted against
time for the free radical and iron signals, as observed by electron

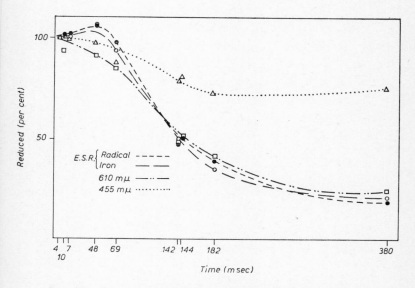

FIG. 5.7. Kinetic measurements on reduction and reoxidation
of dihydroorotic dehydrogenase

The variation in E.S.R. signal strength of the free radical and iron
components is compared with the change in peak height of the
optical absorptions at 610 mμ and 455 mμ. The enzyme is first rapidly
mixed with millimolar DPNH, and then after 30 sec a solution of
orotate, acting as an electron acceptor, is added. The resultant reacting
solution was then deep frozen at the times indicated along the horizontal axis (From Aleman *et al.*[15])

resonance, and also from the peaks of the optical spectra observed
at 610 and 455 mμ. It is clear from these kinetic measurements
that the optical peak of 610 mμ follows the time variation of both
the free radical and the iron electron resonance signals identically,
whereas the 455 mμ peak has quite a different kind of variation.
Although the details of the mechanism in this particular enzyme
still need to be clarified, such comparative kinetic studies as these
make quite clear which parts of the optical and electron resonance
spectra are to be associated with one another.

Another dehydrogenase which has been studied in considerable detail by electron resonance methods[16, 17] is succinic dehydrogenase. This enzyme is, however, difficult to prepare under uniform conditions, and its extreme lability makes it very difficult to be certain what is actually being investigated in electron resonance or other forms of spectroscopy. A signal typical of a free radical is readily observed, however, on addition of substrate to the enzyme, and at low temperatures an additional signal corresponding to iron at a $g = 1.94$ is also obtained. Detailed kinetic studies on this enzyme, such as those carried out for xanthine oxidase, have however still to be made, and in the meantime positive identification of the enzyme mechanism is not too reliable. This is not due to a failing of the electron resonance technique, but an inability to prepare uniform and well characterized specimens for study.

The situation with DPNH dehydrogenases, which are again derived from the respiratory chain of mitochondria, is very similar to that of the succinic dehydrogenase. Thus, they can again be obtained in a variety of different forms, and the electron resonance studies of them have also yielded various different results. The more complex preparations all appear to show the $g = 1.94$ signal on reduction with the DPNH, whereas this signal has not been found in preparations that have had alcohol treatment at low pH values. A large number of different studies have, in fact, been carried out on these enzymes by Beinert and his colleagues,[18] and a recent summary of the various results that have been obtained on these and other related enzymes is given in a review article by Beinert and Palmer.[19]

§5.7

THE $g = 1.94$ SIGNAL ASSOCIATED WITH IRON ATOMS

It will have been noted in the last sections, that reference is often made to an electron resonance signal obtained from iron atoms and occurring at a g-value of 1.94. This signal is, in fact, a fairly common one obtained under reducing conditions in many organisms and preparations; it has its major g-value between 1.93 and 1.95, with a minor component reaching to $g = 2.0$. The fact that the 1.94 and the 2.00 components do represent the genuine g-value spread associated with one electron resonance entity was confirmed by Sands and Beinert,[20] who showed that the two components appeared and disappeared with a constant intensity ratio and must, therefore, correspond to a g-value anisotropy and be associated with an iron atom at a site with non-spherical symmetry. Measurements at 12 and 35 Gc/s also confirmed that the peaks were indeed due to a g-value variation and not to any form of electronic or hyperfine splitting.

More recent work[16, 21] has, in fact, shown that the three principal g-values, g_x, g_y, g_z, can often be resolved, and this would correlate with the suggestion that the signal appears to be associated with an iron atom bound to the protein at a site with partial orthorhombic symmetry. Moreover it has also been shown that the iron of the complex characterized by this g-value of 1·94, cannot be affected by various chelating agents, unless the protein is partly denatured. It would appear, therefore, that the iron complex is intimately involved in the micromolecular structure of the protein itself. Moreover a hyperfine splitting due to the relatively small amount of Fe^{57} isotope can just be detected,[21] and this in turn would suggest that the electrons are relatively closely bound to the iron atom and not moving in delocalized orbitals.

A g-value which averages to below 2·00 is not to be expected for iron in a normal low-spin state; in the next chapter it will be seen that the average for a typical low-spin iron protein, such as metmyoglobin azide, is 2·3. It would therefore appear that these iron atoms are bound in a very specific manner, characteristic of a non-haem metal-protein linkage, but nevertheless of relatively common occurrence in biological systems. A variety of different theories have been put forward to account for this particular signal, and several of them have suggested that this g-value is produced by strong internal electric fields of low symmetry.

Thus, Van Voorst and Hemmerich[23] suggested a model based on reduced nitro-prusside with a strong tetragonal distortion and a not-too-low-lying ligand level. The iron could then share the unpaired electron with one of the ligands and the formal valency of the iron would be undetermined. In a similar way, Brintzinger *et al.*[35] have analysed various cases of complexes having both one and two iron atoms and in which strong tetrahedral distortion leads to g-values below 2·0. It is suggested that in the two-iron complex a diamagnetic compound, or broadened absorption, may be present in the oxidized state, while the addition of one electron to the complex, during the reduction process, produces a single paramagnetic iron atom in each complex and a readily observable signal. Quite a different type of explanation has been put forward by Gibson and Thornley,[22] who suggest that there may be an antiferromagnetic interaction between the two high-spin valency states of the iron atoms, possibly acting via an intervening ligand atom, such as sulphur. One of the iron atoms is assumed to be in a high-spin state with $S_1 = \frac{5}{2}$, while the other has $S_2 = 2$. These thus interact antiferromagnetically to give a ground doublet with spin $\frac{1}{2}$. The $g = 1 \cdot 94$ can then be explained by taking a normal g of $2 \cdot 02$ for the ferric state, and by choosing an appropriate energy level scheme for the ferrous ion with the d_{z^2} orbital lowest, and the d_{yz} next, with splittings which are typical of a tetrahedral distortion.

Various experimental investigations have been made of recent years, to try and decide between these theories. The work of Orme-Johnson, Hansen, Beinert *et al.*,[36] in which complexes enriched with Fe^{57} and S^{33} nuclei have been studied, is a good example of such investigations. In such complexes, the hyperfine structure due to the Fe^{57} can be clearly resolved, and information on the wave functions associated with the iron atoms can then be deduced from the measured splittings. These results can, in fact, be explained only in terms of a two-iron model for the complex, and there seems definite evidence for an interaction of the sulphur atoms with the electrons responsible for the $g = 1 \cdot 94$ signal.

It would probably be fair to say that the exact reason why this particular g-value is obtained from these non-haem iron proteins is not yet fully understood, and more experiments must be undertaken before a complete explanation is obtained and the detailed nature of the complex itself is resolved. Nevertheless, the signals can be used very effectively at the moment as a monitor of the amount of this particular protein-bound iron present in biological systems.

§5.8

STUDIES ON CATALASE AND PEROXIDASE

The two enzymes catalase and peroxidase occur very generally, catalase being found in practically all forms of life, both animal and plant, while peroxidase occurs in nearly all plant cells. They are probably two of the most important enzymes that exist, and they are very similar in that they both have the same prosthetic group, based on the haem or porphyrin plane. They are, therefore, considered together in this section, although neither of them is an oxidative enzyme, as such, and they really fall into two different groups, catalase being a *splitting enzyme*, whereas peroxidase can be classified as a *transferring enzyme*.

Thus the essential action of catalase is to bring about the decomposition of hydrogen peroxide into water and oxygen, whereas peroxidase has no decomposing action on pure hydrogen peroxide; but, if a suitable acceptor molecule is present, it will catalyse the transfer of oxygen to the acceptor which itself therefore becomes oxidized. The function of the catalase enzyme is to protect the living organism from too high a concentration of hydrogen peroxide, this being the product of the action of various oxidases in the system. It can be fairly readily isolated from such sources as beef liver and can be obtained in a pure crystallized form. It is highly specific with respect to its reaction with hydrogen peroxide, since it does not catalyse the decomposition of any other peroxides, whereas its catalyctic activity with the hydrogen peroxide is extremely efficient. Thus at $0°C$, one molecule of catalase can produce the decomposition of five million molecules of hydrogen peroxide in a time of 1 minute, and this turnover rate is one of the highest that has ever been determined.

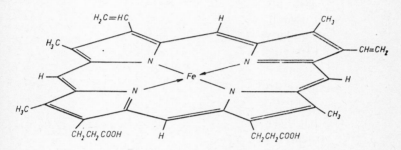

FIG. 5.8. Structural diagram of haem plane

The iron atom is at the centre of the haem plane and is surrounded by four nitrogens, as shown. Different groups are attached to the fifth and sixth coordination points of the iron atom, as explained in the next chapter.

The active part of the molecule is linked with its prosthetic group of the haem plane, which is shown in Fig. 5.8, and it can be seen that this contains a central iron atom in the middle of a square of four nitrogen atoms. It is therefore to be expected that the enzymatic activity of the catalase will probably be connected with valency changes in this central iron atom, and hence electron resonance studies on its action might be expected to show both free radical signals and possibly some from various states of the iron atom. The structure of peroxidase is very similar in this respect, since its active prosthetic group is also the haem plane, and again earlier chemical studies had suggested that the enzyme's substrate complex, formed during the enzymatic activity of the peroxidase, would be associated with valency changes in the iron atom.

Since both of these enzymes can be prepared in a crystalline state of high purity, they have been studied by various techniques for some time, although standard chemical methods of investigation had not been able to elucidate a completely acceptable picture of the mechanism of the enzyme's operation. As a result of very systematic studies, George[24] proposed a scheme in which two intermediate compounds were formed during the enzymatic action, thus:

$$\text{Peroxidase} + H_2O_2 \rightarrow \text{Compound I}$$
$$\text{Compound I} + AH_2 \rightarrow \text{Compound II} + AH^\cdot$$
$$\text{Compound II} + AH_2 \rightarrow \text{Peroxidase} + AH^\cdot$$

This kind of mechanism therefore suggests two one-electron reduction steps, and predicts that there will be intermediates formed as free radicals, $AH^\cdot$. It is very hard to produce definite proof for these intermediates by normal chemical methods, but this is just the kind of situation in which electron resonance studies should be able to produce conclusive evidence, one way or the other.

The study of such peroxidase systems was in fact one of the first successful detailed investigations of enzyme activity by electron resonance, and the work has been undertaken systematically by Yamazaki, Piette and Mason.[25, 26] The main object of these studies was to search for the intermediate, $AH^\cdot$, and determine the conditions under which such a free radical signal would occur. Initially rapid freezing techniques were employed, a reacting mixture of the solvents being suddenly deep frozen after an appropriate time interval from initiation. These first experiments were mainly of a negative nature, however, and although at the time it was thought that this might have been due to the low concentration of radicals present, it was probably due rather to the line broadening that occurred in the solid specimens owing to the dipolar interaction, which was not being averaged out by the tumbling motion of the molecules in the liquid state. Further experiments were, therefore, carried out at room temperature, and it was these particular studies that first showed the power of the continuous flow and stop-flow techniques in such enzyme studies. Yamazaki, Piette and Mason[25, 26] were able to trace out the complete spectrum observed at different time intervals of reaction from the mixing chamber by employing the continuous flow method described in §3.4.2, and an example of some of the spectra thus obtained is shown in Fig. 5.9. Enzyme concentrations of 10^{-8}–10^{-7} molar were employed, with substrate concentrations of 2×10^{-2} to 5×10^{-3} molar, and these produced steady-state free radical concentrations of about 10^{-6} molar. One of the advantages of these early studies was that the enzyme being used could be obtained in fairly large quantities, and hence the considerable consumption of

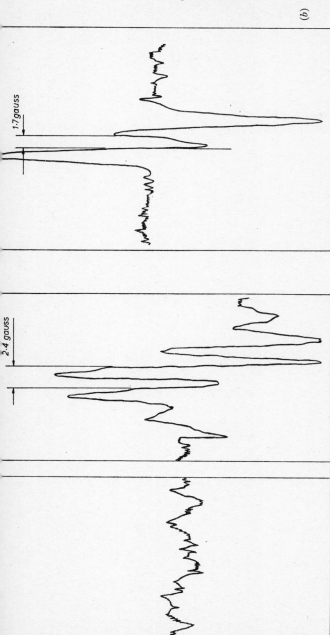

FIG. 5.9. Free radicals produced in continuous-flow studies on peroxidase

The left set of spectra (a) show the free radicals obtained when peroxidase catalyses the oxidation of hydroquinone. No signal is observed in the absence of the enzyme, on the left, whereas a well-resolved five line hyperfine structure is obtained from the benzosemiquinone radicals on the right. The right-hand spectrum (b) shows similarly the effect of peroxidase on the oxidation of ascorbic acid. In this case the free radicals produce a doublet hyperfine pattern (From Yamazaki *et al.*[25])

material presented by these figures was not a serious drawback to the experiments.

A variety of experiments employing different substrates were carried out and the results shown in Fig. 5.9 are those obtained on the one hand with hydroquinone and hydrogen peroxide, and on the other hand with ascorbic acid and hydrogen peroxide. It is evident that in both cases a significant free radical signal is only obtained when the peroxidase enzyme is also present to catalyse the reaction, and that when this is so, a large free radical concentration can be measured and identified from the resultant hyperfine structure. Thus the five-line spectrum of Fig. 5.9(*a*), can be readily identified with the benzosemiquinone free radical, as explained in detail in §1.5, and already given as a spectrum by way of example in Fig. 1.13. The two-line spectrum of Fig. 5.9(*b*), however, is that expected from the ascorbic acid radical, both free radicals being present in the anionic form at the *p*H values used in these experiments.

These continuous-flow experiments therefore demonstrated immediately that there were indeed free radical species associated with the intermediates of the peroxidase activity, and a systematic study of the kinetics of the reaction was then undertaken by varying the time between initial mixing of the solutions and the measurement in the electron resonance cavity. In this way, they were able to show that the steady-state concentrations of the free radicals were proportional to the square root of the total enzyme concentration, and this relationship was taken to support the mechanism previously suggested by George,[24] and in particular the fact that neither compound I nor compound II is reduced by the free radical intermediates themselves, but that the free radicals decay either by dismutation or by dimerization. More recently, direct evidence for the decay by dismutation has been obtained from measurements by Piette *et al.*,[27] who used a combination of optical and electron resonance flow techniques to study the oxidation of the drug chlorpromazine in a peroxidase–hydrogen–peroxide system. The chlorpromazine free radical which is formed is in fact very stable and could be quantitatively accumulated at *p*H 4·8. It could then be induced to decay by dismutation, by raising the *p*H value, and a rate constant for this reaction could thus be measured directly. Moreover, in the conditions of low *p*H, and therefore no radical dismutation, the reaction

$$\text{AH·} + \text{peroxidase compound II} \rightarrow \text{A} + \text{peroxidase}$$

becomes measurable by direct kinetic studies, whereas this had not been possible in the earlier investigations. A direct plot of the kinetics of this reaction for two different concentrations of peroxidase is shown in Fig. 5.10 and provides a very elegant example of how the kinetics of such enzyme reactions can be separated into first and second stages and each stage then measured quantitatively on its own.

As well as these detailed studies on the kinetics of the free radicals associated with the peroxidase activity, E.S.R. measurements have also been made on the iron atoms in various peroxi-

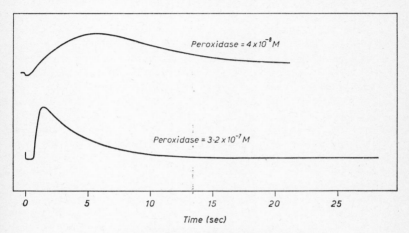

Time (sec)

FIG. 5.10. Kinetics of oxidation of chlorpromazine
in the presence of peroxidase

Both curves were obtained for the same concentrations of chlorpromazine and hydrogen peroxide and at a pH of 4·8. The upper curve has a concentration of 4×10^{-8} M peroxidase, while the lower curve has a concentration of $3·2 \times 10^{-7}$ M, and gives a very much shorter reaction time (From Piette *et. al.*[27])

dases. These results are summarized at the end of §6.6.5 in the next chapter, after the various types of E.S.R. signal that can be obtained from haem-bound iron have been discussed.

§5.9

STUDIES ON ENZYMES CONTAINING COPPER

A great many electron resonance studies have been undertaken on enzymes containing copper atoms. These not only comprise some of the most interesting enzymes biochemically, but the well-resolved hyperfine pattern, obtained from the copper ions, also allows a considerable amount of information to be obtained on the chemical bonding of the copper atoms from the electron resonance spectra themselves. The way in which the actual magnitudes of the *g*-values, and hyperfine splittings, can be used to deduce details of the chemical bonding, and molecular structure, are considered more fully in the next chapter. This kind of analysis of the E.S.R. spectra of enzymes will therefore be left until the more detailed theory has been developed, and it will be seen in §6.4 that very precise information on the coordination and bonding of copper in such enzymes as ceruloplasmin has been obtained in this way.

Ceruloplasmin is a blue-coloured protein with a molecular weight of 160 000 and contains eight copper atoms per molecule. Moreover, differential chemical analysis[28] shows that more than 36 per cent of the copper is in a cupric valence state, and it is probable that in fact there are equal amounts of cuprous and cupric copper in such copper proteins. The detailed analysis of the *g*-values and hyperfine splittings observed in ceruloplasmin, summarized in §6.4, shows that there is strong evidence for different kinds of bonding amongst the four cupric ions, but it is of interest that normal kinetic studies on such enzymes also help in the elucidation of the contribution that the different valence states of copper make to its activity.

Thus, for example, several attempts have been made to correlate the blue coloration of the enzyme with both its enzymatic activity and its E.S.R. spectrum. Broman *et al.*[29] found that, on incubation of the enzyme, a very close relationship existed between the disappearance of the 610 mμ absorption, the enzyme activity, and the amplitude of one of the low field E.S.R. lines. In the same way, Blumberg *et al.*[30] were able to show that there was a linear relationship between the integrated intensity of the E.S.R. absorption and the bleaching of the 610 mμ line when the enzyme was reduced by ascorbate.

It therefore seems clear that the copper not only has a definite role in the enzymatic activity, but is also probably responsible for the blue coloration of the protein, either via a cuprous-cupric interaction like the charge-transfer band in Russian blue, or possibly via charge-transfer transitions from nitrogen to copper.

Another enzyme which contains copper and has been studied fairly extensively by electron resonance techniques is cytochrome oxidase. The cytochromes themselves form a very important group of substances concerned in the later stages of many biological oxidations. Like catalase and peroxidase they also contain the haem group as an essential constituent, and various photo-chemical studies show that the iron atom of this haem group takes part in the oxidation process itself. Evidence also accumulated,[31] however, that copper as well as these iron atoms was involved in the bio-synthesis of the enzyme. Cytochrome oxidase itself is the enzyme which plays the final role in these oxidative systems, in that it brings about the oxidation by molecular oxygen of the reduced forms of the cytochromes. When purified preparations of the cyto-chrome oxidase became available, detailed chemical analysis showed that there was a definite ratio of the copper present per molecule of haem in the enzyme, and early chemical methods used to investigate this copper suggested that it was in a cuprous form. Other measurements, however, suggested that sometimes the copper could be obtained in a cupric state, and more definite results on the actual role of copper in the enzyme were not available until the advent of the electron resonance technique.

There are several unusual features about the electron resonance spectrum [32,33] observed from the copper in the cytochrome oxi-dase and they can be used to give some information on the binding of the copper within this molecule. Thus the g-values of $g_{\parallel} = 2 \cdot 17$ and $g_{\perp} = 2 \cdot 03$ are very much closer to the free electron value than those normally observed for copper in other copper-containing proteins. As explained in more detail in the next chapter, this sug-gests that there must be very strong covalent bonding between the copper atoms and the other ligands, since a large reduction of the g-value divergence like this corresponds to a significant reduction in the value of the constant α^2 in equation (5.1) and hence to a very high degree of covalency. This conclusion is also supported by the fact that there is no observable hyperfine splitting around the $g_{\parallel}$ direction, and that therefore such splitting must be of a very small value, again confirming high delocalization of the unpaired electron from the copper atom. Extensive studies on this copper spectrum have also been made by relaxation methods, [34] details of which are described in Chapter 7. It was initially thought that these results indicated a strong exchange interaction between pairs of copper atoms, since no saturation effect of the absorption could be observed even at 93°K with high microwave power levels. More recent measurements at liquid helium temperatures have shown, however, that such saturation does take place at low power intensi-ties, and therefore strong exchange interaction between the copper atoms is probably unlikely.

Various studies on the kinetics of the enzyme activity and the copper signal have also been made, and in summary it can be said that these electron resonance studies have confirmed that the copper found in cytochrome oxidase is a real component of the enzyme and able to undergo rapid oxidation and reduction. It is also clear, however, that considerably more work needs to be done to elucidate the exact nature of the bonding of the copper atoms and to determine whether any copper-haem interaction takes place.

This example of an important enzyme, in the investigation of which electron resonance has already made some important contributions and can be expected to make many more, may, indeed, form an appropriate topic on which to end this chapter on enzymes. Most of the enzymes are very complex molecules, and very few have had their structure or mechanism completely elucidated as yet. They therefore offer a particularly interesting challenge to the methods and techniques of electron resonance, and it is hoped that the examples given in the preceding pages will give some indication of the way in which electron resonance can help, both in the detailed study of the kinetics and mechanism of the enzyme activity and in the investigation of the precise way in which the active metal ions are bound to the rest of the protein.

References

1. Assenheim, H. M., *Introduction to Electron Spin Resonance*, Chap. 2 (Adam Hilger, London, 1966; Plenum, N.Y.);
   Orgel, L. E., *Ligand Field Theory*, (Methuen, London, 1960);
   Griffiths, J. S., *Theory of Transition Metal Ions* (C.U.P., London, 1961).
2. Ingram, D. J. E., *Spectroscopy at Radio and Microwave Frequencies*, p. 171 (Butterworths, London, 1967).
3. Deal, R. M., Ingram, D. J. E., and Srinivasan, R., *Proc. XII Colloq. Ampère*, p. 239 (1963).
4. Windle, J. J., Wiersema, A. K., Clark, J. R., and Feeney, R. E., *Biochemistry*, 1963, **2**, 1341.
5. Michaelis, L., and Menten, M. L., *Biochem. Z.*, 1913, **49**, 333.
6. Bray, R. C., Malmstrom, B. G., and Vanngard, T., *Biochem. J.*, 1959, **73**, 193.
7. Bray, R. C., *Biochem. J.*, 1961, **81**, 196.
8. Bray, R. C., Pattersson, R., and Ehrenberg, A., *Biochem. J.*, 1961, **81**, 178.
9. Bray, R. C., Palmer, G., and Beinert, H., *J. Biol. Chem.*, 1964, **239**, 2667.
10. Palmer, G., Bray, R. C., and Beinert, H., *J. Biol. Chem.*, 1964, **239**, 2657.
11. Bray, R. C., and Meriwether, L. S., *Nature*, 1966, **212**, 467.
12. Rajagopalan, K. V., and Handler, P., *J. Biol. Chem.*, 1964, **239**, 2022.
13. Beinert, H., and Palmer, G., *Advances in Enzymology*, 1965, **27**, 168.
14. Palmer, G., and Beinert, H., *Rapid Mixing and Sampling Techniques in Biochemistry*, p. 205 (Academic Press, New York, 1964).
15. Aleman, V. A., Handler, P., Palmer, G., and Beinert, H., *J. Biol. Chem.*, 1967, **243**, 2560.
16. King, T. E., Howard, R. L., and Mason, H. S., *Biochem. Biophys. Res. Comm.*, 1961, **5**, 329.
17. Hollocher, T. C., and Commoner, B., *Proc. Nat. Acad. Sci.*, 1961, **47**, 1355.
18. Beinert, H., Palmer, G., Cremona, T., and Singer, T. P., *Biochem. Biophys. Res. Comm.*, 1963, **12**, 432.
19. Beinert, H., and Palmer, G., *Advances in Enzymology*, 1965, **27**, 177.
20. Sands, R. H., and Beinert, H., *Biochem. Biophys. Res. Comm.*, 1960, **3**, 41, 47.
21. Shethna, Y. I., Wilson, P. W., Hansen, R. E., and Beinert, H., *Proc. Natl. Acad. Sci.*, 1964, **52**, 1263.

22. Gibson, J. F., Hall, D. O., Thornley, J. H. M., and Whatley, F. R., *Magnetic Resonance in Biological Systems*, p. 181 (Pergamon Press, London, 1966, and *Proc. Nat. Acad. Sci.*, 1966, **56**, 987.

23. Van Voorst, J. D. W., and Hemmerich, P., *Magnetic Resonance in Biological Systems*, p. 183 (Pergamon Press, London, 1966).

24. George, P., *Nature*, 1952, **169**, 612; and *Biochem. J.*, 1953, **54**, 267.

25. Yamazaki, I., Mason, H. S., and Piette, L. H., *J. Biol. Chem.*, 1960, **235**, 2444.

26. Yamazaki, I., and Piette, L. H., *Biochim. Biophys. Acta.*, 1961, **50**, 62 and 1963, **77**, 47.

27. Piette, L. H., Bulow, G., and Yamazaki, I., *Biochim. Biophys. Acta.*, 1964, **88**, 120.

28. Kasper, C. B., Deutsch, H. F., and Beinert, H., *J. Biol. Chem.*, 1963, **238**, 2338.

29. Broman, L., Malmstrom, B. G., Aasa, R., and Vanngard, T., *J. Mol. Biol.*, 1962, **5**, 301.

30. Blumberg, W. E., Eisinger, J., Aisen, P., Morrell, A. G., and Scheinberg, I. H., *J. Biol. Chem.*, 1963, **238**, 1675.

31. Wainio, W. W., Van der Wender, C., and Shimp, N. F., *J. Biol. Chem.*, 1950, **234**, 2433.

32. Beinert, H., Griffiths, D. E., Wharton, D. C., and Sands, R. H., *J. Biol. Chem.*, 1962, **237**, 2337.

33. Sands, R. H., and Beinert, H., *Biochem. Biophys. Res. Commun.*, 1959, **1**, 175, and 1960, **3**, 47.

34. Beinert, H., and Palmer, G., *J. Biol. Chem.*, 1962, **237**, 3843, and 1964, **239**, 1221.

35. Brintzinger, H., Palmer, G., and Sands, R. H., *Proc. Nat. Acad. Sci.*, 1966, **55**, 397.

36. Orme-Johnson, W. H., Hansen, R. E., Beinert, H., Tsibris, J. C., Tsai, R. L., and Gunsalus, I. C., *Proc. Nat. Acad. Sci.*, 1968.

Chapter 6

The Investigation of
Metallo-Organic Compounds

§6.1

TRANSITION GROUP ATOMS AND THEIR
ROLE IN BIOCHEMISTRY

It has already been pointed out that one of the great advantages of the electron resonance technique is that it can give very detailed information on both the position and site-symmetry of transition group atoms and also on their detailed chemical bonding to the surrounding ligand atoms. This information can be obtained even when the molecules are very large and contain 10 000 other atoms or more, since all the diamagnetic atoms are completely ignored by the electron resonance technique. This particular feature of being able to isolate and study in detail one particular atom is of great importance in biochemistry, where the interesting molecules are often very complex and such techniques as infra-red spectroscopy or X-ray crystallography give extremely complex spectra, with absorption lines overlapping from thousands of different atoms or groups.

It can, of course, be argued on the other side that the technique of electron resonance is very limited, since it can give information on only one particular atom, but it often transpires that this transition group atom itself is one of real interest in the molecule and is often partially or wholly responsible for the specific activity of the molecule concerned. This is not surprising, since the very property which gives the transition group atom its magnetic characteristics, i.e. the presence of the unpaired electron in its orbital, is also often the property which gives the specific chemical activity to this particular atom or group. Hence the electron resonance technique is often able to study specifically the actual unpaired electrons which are responsible for the valency changes, or catalytic activity, associated with the biochemical molecule as a whole.

This fact has already been well illustrated in the previous chapter, where various electron resonance studies on enzymes have been considered. It became clear there that, in practically every case where the enzyme contained a transition group atom in its molecule, this atom took a direct role in the oxidation, or valency change, associated with the enzyme activity. Thus the electron resonance technique was able to probe these changes directly, as well as monitoring any free radical reactions which were taking place at the same time. Enzymes, however, are not the only molecules of biochemical interest which contain transition group atoms, and in this chapter further consideration is given to other types of biochemical molecules in which the transition group atom can be studied in detail and in which these studies can give very important information either on the structure or on the mechanism of the biochemical activity concerned. Most of the transition group atoms that have been studied in this connection are those that have already been considered as taking part in enzyme activity, such as copper and iron, but before specific examples of such studies are considered, a more general account of how the energy levels of such transition group atoms are affected by their molecular environment will be given, so that the general features of such studies can be more readily appreciated.

§6.2

ATOMIC ORBITALS AND ENERGY LEVELS ASSOCIATED
WITH TRANSITION GROUP ATOMS

It has already been noted in the introductory section that the feature distinguishing the transition group atoms from the free radicals is the location of their unpaired electrons mainly in atomic orbitals closely linked with one particular atom, and such orbitals therefore have strong coupling to both the spin and orbital motion of the atom as a whole. As a result of this, the g-value associated with the electron resonance often departs widely from that of the free spin, and a measurement of this g-value can be used to give detailed information on the nature of the chemical bonding of the atom with which the electron is associated. If such a transition group atom is completely free, and does not experience any electrostatic fields from either the surrounding crystal lattice or other atoms in the molecule, then the g-value of the unpaired electron would be given by the expression for the Landé splitting factor, as has already been quoted in equation (1.10). Electron resonance studies on such free atoms[1, 2] obtainable in a monatomic vapour have confirmed that the g-values then measured are those given by this expression. In the vast majority of cases in physics, chemistry and biochemistry, however, the transition group atoms of interest are not in a free state, but are firmly bound either within a molecule or within a crystal lattice. There are then strong electrostatic fields acting on these atoms, which couple to the electron orbital motion in particular, and these fields can cause the g-values to shift very appreciably from the simple expression given for the case of the free atom.

Although the full theory of how such g-value shifts can be correlated to the magnitudes and symmetry of the internal electrostatic fields would not be appropriate in a book of this type, some general qualitative understanding of the interactions that are taking place is necessary, if the implications of the measurements on metallo-organic compounds in biochemistry are to be appreciated. The situation is probably best approached by considering the actual spatial orientation of the electron wave function normally associated with these transition group atoms, and seeing how the internal molecular fields will then affect the energies associated with these spatial orientations. Most of the transition group atoms that have been studied so far belong to the first transition group, and therefore their electrons will be moving in $3d$ orbitals. The detailed consideration of how these orbitals are affected by the internal molecular fields will therefore be confined to a discussion of the $3d$ orbitals in the first place. Again it is probably helpful to begin with a discussion of the orbitals associated with the free atom, and these are illustrated in Fig. 6.1. It is seen that there are, in fact, five possible $3d$ orbitals for such an unpaired electron, and these can be characterized by the way in which their lobes are orientated with respect to the three mutually perpendicular axes, x, y, and z. Thus the first orbital, labelled d_{xy}, has four lobes which intersect the x and y axes respectively. In a similar way, the d_{yz} and d_{zx} orbitals have the same four lobes, but these are now orientated so as to bisect the yz and zx axes respectively. A fourth orbital is then obtained by a rotation of the d_{xy} orbital in the xy plane, so that the four lobes are now pointing along the x and y axes, as shown in Fig. 6.1(d), while the last and fifth orbital, as shown in Fig. 6.1(e), has two lobes pointing along the z axis, together with an annulus of electron density in the xy plane itself.

It is evident that four of these orbitals are in fact identical in spatial variation, and only their orientation with respect to the axes alters. The reason why the fifth one is different in form, is because this particular axis, Oz, is the one that has been chosen as the axis of quantization, and the electron distribution of all five orbitals, when summed together, must produce spherical symmetry, corresponding to a filled closed shell of electrons. In the absence of any external electric or magnetic field, there is no direction with which

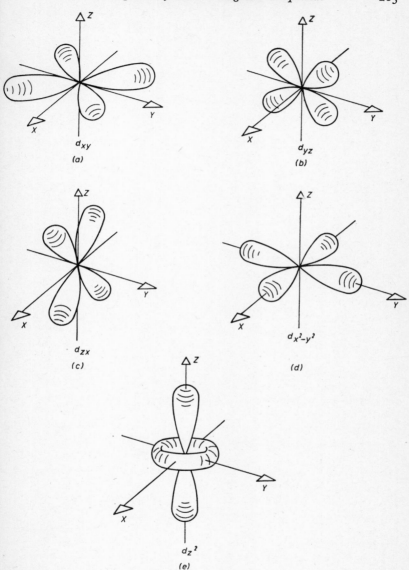

Fig. 6.1. 3d orbitals of free atom

The spatial orientations of the five independent 3d orbitals are shown as referred to three mutually perpendicular axes. Cases (a), (b) and (c) intersect these axes, while (d) and (e) point along them

any of three axes could be identified, and hence it follows that, for the free atom, all of these five orbitals are degenerate, i.e. they all have the same energy. If, however, an external electric or magnetic field is applied in a definite direction, it is then possible for these

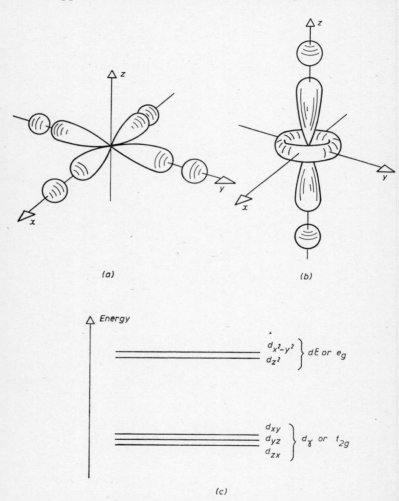

FIG. 6.2. Transition group atom surrounded by octahedron of ligand atoms

The interaction between the various $3d$ orbitals and the electrons on the ligand atoms is shown (a) for the $d_{x^2-y^2}$ orbital and the four ligand atoms along the x and y axes; and (b) for the d_{z^2} orbital and the two ligand atoms along the z axis. The resultant shift in the energies of these orbitals is shown in (c)

five orbitals to take up specific orientations with respect to the field and also to have their energies either raised or lowered. The way in which this is done can be illustrated by taking the case of such a transition group atom surrounded by six ligand atoms in an octo-hedron, with the atoms arranged in pairs at equal distances along the positive and negative x, y and z axes, as illustrated in Fig. 6.2.

In the first approximation, the electron distribution around these ligand atoms may be considered as a sphere of electrical charge, and it now becomes obvious that there will be considerably more repulsion between the electron density on the six ligand atoms, and the electron distribution in the $d_{x^2-y^2}$ and in the d_{z^2} orbitals, than in the other three. Thus the four lobes of the $d_{x^2-y^2}$ orbital point directly at the four atoms along the x and y axes respectively, and hence a noticeable repulsion takes place between them; and in the same way, the two lobes of the d_{z^2} orbital point directly at the two atoms along the z axis. More energy will therefore be required to place electrons in these two orbitals than in the other three orbitals, since in their cases the lobes of electron density point away from the positions of the ligand atoms, and hence no repulsion takes place. In other words the five orbitals now split into two groups. One group contains the $d_{z^2-y^2}$ and the d_{z^2} orbital, which is often called the dε, or the e_g group, and is raised in energy. The other group contains the d_{xy}, d_{yz} and d_{zx} orbitals, which are often referred to as the dγ, or the t_{2g} group, and this is lowered in energy. This overall splitting of the energy levels is shown in Fig. 6.2(c). Separation of the d-orbitals into these two groups by the electro-static field, produced by such an octahedral symmetry of the sur-rounding ligand atoms, is one of the basic effects produced by such a molecular environment. If the surrounding ligands had not had an octahedral symmetry, but rather a tetrahedral symmetry, as is present in such structures as the diamond lattice, then the effect on the two groups of orbitals would have been inverted. Thus the surrounding ligands would be in positions to which the d_{xy}, d_{yz} and d_{zx} orbitals were pointing, while the $d_{x^2-y^2}$ and d_{z^2} orbitals would be pointing in directions of zero ligand spin density.

In all practical cases within a crystal or molecule, the field sym-metry tends to be somewhat distorted from either pure octahedral or pure tetrahedral symmetry, however, and the effect of such distortions can again be readily seen in terms of a slight movement of one pair of the ligand atoms. Thus in Fig. 6.3(a), the octahedral symmetry is shown somewhat distorted by a compression of the two ligand atoms along the Oz axis, so that these are now slightly closer to the central transition group atom than the four ligand atoms along the x and y axes. It is now clear that the repulsion

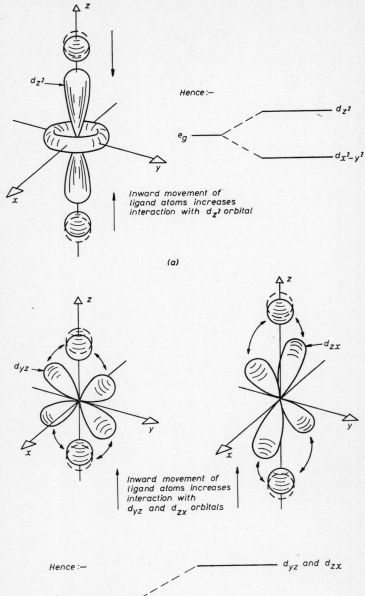

d_{z^2}

Hence :−

e_g d_{z^2}

$d_{x^2-y^2}$

Inward movement of ligand atoms increases interaction with d_{z^2} orbital

(a)

d_{yz}

d_{zx}

Inward movement of ligand atoms increases interaction with d_{yz} and d_{zx} orbitals

Hence :−

t_{2g} d_{yz} and d_{zx}

d_{xy}

(b)

between the d_{z^2} orbitals and these two ligand atoms along the z axis will be even more pronounced than before, and hence the d_{z^2} orbital will be raised to an even higher energy. On the other hand, there will be no additional repulsion between the lobes of the $d_{x^2-y^2}$ orbital, and the four ligand atoms in the xy plane, and therefore this will not be raised in energy in the same way. Hence this additional tetragonal distortion of the octahedral field symmetry will now split the two levels within the upper e_g group. In the same way, the decrease in distance between the two ligand atoms along the z axis will produce a somewhat greater repulsion between them and the lobes of the d_{yz} and d_{zx} orbitals, whereas the four lobes of the d_{xy} orbital will again not be affected. In other words, an energy level splitting is also produced between the orbitals in the t_{2g} group, and this illustrates the general feature that lower symmetry electrostatic molecular fields will tend to produce further splittings in the energies of the orbital groups.

So far, the electron density around the ligand atoms has been considered as a simple sphere, whereas, in most organic compounds, π bonding as well as σ bonding is often present and hence the electron orbitals on the ligand atoms often have noticeable electron density in the p-orbitals, as well as in the spherically symmetrical s-orbitals. The effect of such p-orbital distributions should therefore also be considered, and this is now illustrated in Fig. 6.4. Again the case of the six surrounding ligands equally spaced along the x, y and z axes can be considered, and the p-orbital distribution on the surrounding ligand atoms is shown by the two lobes drawn around each such atom. It is assumed in this case that the xy plane now corresponds to a molecular plane in the structure being considered, as for example the haem or porphyrin plane in the cytochrome, catalase, peroxidase or haemoglobin molecules. In this case the p-orbitals will be pointing at right angles to the xy plane, parallel to the Oz axis, and the way in which these affect the energies of the three orbitals in the t_{2g} group can be seen from the figure. It is seen in Fig. 6.4(a) that the lobes of the d_{xy} orbital, which are confined in the xy plane, point in directions which avoid the p-lobes on the surrounding ligand atoms, and therefore the energy of this particular level will not be changed by such p-orbitals or π bonding. On the other hand, the lobes of the

---

Fig. 6.3. (*facing*) Effect of tetragonal distortion on octahedral symmetry
A tetragonal distortion is produced by moving the two ligand atoms along the z axis closer together. The resultant additional interactions with the different d orbitals can then be seen. In (a) the additional interaction between the two shifted atoms and the d_{z^2} orbital is shown, and in (b) the interaction with the d_{yz} and d_{zx} orbitals

d_{yz} orbital now approach quite close to the p-orbitals on the ligand atoms along the y axis, as shown in Fig. 6.4(b), and hence additional repulsion takes place between these, and the energy of the d_{yz}

(a)   (b)

FIG. 6.4. Interaction with p-orbitals of the ligand atoms in the molecular plane
The interaction of the d_{xy} and d_{yz} orbitals with the p orbitals on the four surrounding ligand atoms, is shown. In (a) it is seen that the d_{xy} orbital is located in the null plane of the p orbitals, while in (b) additional interaction takes place between the p orbitals and the d_{yz} orbital

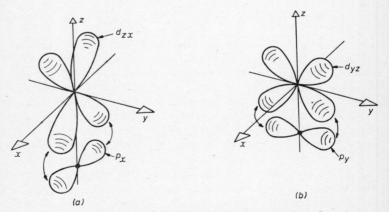

(a)   (b)

FIG. 6.5. Interaction with p orbitals of ligand atoms along Oz axis
It is seen that there is now an additional interaction with either the d_{zx} or d_{yz} orbital, depending on whether the p orbital of the ligand atom is parallel to the Ox axis, as in (a), or the Oy axis, as in (b)

orbital is thus raised. The same comments apply equally to the lobes of the d_{zx} orbital, and hence the effect of these p-electrons of the four atoms in the xy plane is again to separate the t_{2g} group of levels into a lower d_{xy} and an upper d_{zx} and d_{yz}, which will still remain together.

The interaction with the p lobes on the two ligand atoms along the z axis may now be considered, and the orientation of these p-orbits will be determined by structural features of the molecule outside the xy molecular plane. The two cases in which the p-orbit is parallel to the x and y axes are shown separately in Fig. 6.5(a) and (b) and it is clear from this figure that in the first case the d_{zx} orbital will suffer further repulsion from the proximity of its lobes to the p-orbital parallel to the x axis. If, on the other hand, the p-orbital on the ligand atom is parallel to the y axis, as shown in Fig. 6.5(b), it will be the d_{yz} orbital which suffers further repulsion and is thus raised further in energy. It follows from this qualitative analysis that the actual order of the energy levels within the t_{2g} group is determined by the direction in which the p orbital on the ligand atom along the Oz axis is orientated. Hence a knowledge of this energy level order, within the t_{2g} group, can in turn often be used to give precise structural information about the molecule itself. This point is, in fact, taken up in detail when the electron resonance studies on haemoglobin are discussed.

§6.3

g-VALUES AND MOLECULAR FIELD SPLITTINGS

The way in which the experimentally determined g-values are related to the energy level splittings produced by the molecular field, which have been discussed in the last section, can now be considered. It has been seen that the effect of the internal crystalline, or molecular, field on the orbitals of the metal atom is to remove their degeneracy in energy and produce energy splittings between them, so that, in general, one specific orbital level is left as the ground state. If the splitting between this ground state level and the next highest is very large, and there is no coupling between these levels, the orbital contribution to the electron's magnetization can be considered as completely 'quenched', and the g-value for the ground state is then that for the free electron spin alone. In fact, however, there is still often some significant coupling between the ground-state level and the higher orbital levels, and this is brought about by the spin-orbit coupling parameter already mentioned in the previous chapter. The amount of coupling between the higher levels and the ground state will vary with the ratio of λ/Δ, as previously mentioned in §5.2.1. It will be remembered that, in this expression, λ is the spin-orbit coupling parameter and Δ is the energy level splitting between the ground state and the appropriate higher orbital level. It is evident that the value of Δ, and hence the amount of admixture between the orbital levels, will depend both on the symmetry and the magnitude of the internal molecular electrostatic field.

The effect of this interaction between the orbitals is to admix some of the quantum characterization of the higher orbital levels into the ground state, so that this ground state is no longer described accurately by a single quantum number. Instead, it takes on a characterization which includes a small admixture of other quantum designations. The actual amount and nature of this admixture is determined by applying quantum mechanical perturbation theory, in which the splitting produced by the internal electric field is considered as the first-order effect, and the interaction produced by the spin-orbit coupling is then treated as a perturbing effect on the initial splittings. The detailed way in which this calculation is carried out may be found in standard texts on this subject,[3] but the case of the copper Cu^{++} ion in an octahedral field may be taken by way of illustration.

The cupric ion contains nine electrons in its $3d$ shell, and since the full complement is ten, it follows that there is, in fact, one unpaired electron in this shell, and the single missing electron can thus be treated as a positive hole. The energy level system for this effective positive hole will then be exactly the same as that for a single electron in an octahedral situation, as already considered in some detail and illustrated in Fig. 6.2, but with the energy level completely inverted to allow for the change in sign. In this particular case, the d_{z^2} orbital is thus left as the ground state, with a small splitting to the $d_{x^2 - y^2}$ orbital, and a much larger splitting to the d_{xy}, d_{yz} and d_{zx} group of orbitals. In the normal quantum number designation of M_L and M_S, this ground state would then be classified as $|\, 0, \pm\tfrac{1}{2}\rangle$. The application of general quantum mechanical theorems to the admixing produced by the spin-orbit interaction perturbation shows that the only higher orbitals to be coupled into the ground state will have quantum designations of $|\, 1, \mp\tfrac{1}{2}\rangle$. Moreover, the calculation shows that the value of the admixture coefficient itself is equal to $-(3/\sqrt{2})\lambda/\Delta$, and hence the full designation of the ground state with the admixture produced by the spin orbit perturbation now becomes

$$|\, 0, \pm\tfrac{1}{2}\,\rangle - \frac{3}{\sqrt{2}}\cdot\frac{\lambda}{\Delta}\cdot|\, 1, \mp\tfrac{1}{2}\,\rangle \qquad (6.1)$$

The relevance of this to the g-value calculation is that, by definition, the g-value is the rate at which the two levels still left in this ground state diverge as an external magnetic field is applied to them. Thus, the final theoretical calculation of the g-value is obtained by applying the interaction of the magnetic field to these energy levels and treating the result as a further perturbation on the total system. The direction in which the magnetic field is applied will also be important and different perturbation effects will be obtained in different directions, and this is, of course, the origin of the anisotropy in the g-value itself.

Again, as a specific example, the application of a magnetic field along the Ox direction, in the above case can be considered. This will split the degenerate levels represented by the $| \pm\frac{1}{2} \rangle$ in the M_S quantum number, and they will diverge as the value of the field increases, and the actual energy levels will have the values of $\beta H_x \left(1 - 3 \; \lambda/\Delta\right)$ and $-\beta H_x \left(1 - 3 \; \lambda/\Delta\right)$. Thus an actual energy separation between the two levels is produced, which is equal to $2\beta H_x \left(1 - 3 \; \lambda/\Delta\right)$. In other words, the transition now corresponds to a g-value which may be written

$$g_\perp = 2 \left(1 - 3 \; \frac{\lambda}{\Delta}\right) \tag{6.2}$$

This rather shortened summary of how the magnitude of the g-value is related to both the value of the spin orbit coupling parameter and the value of the splitting produced in the orbital levels by the molecular field may serve to amplify the comments made in a previous chapter and also help in an understanding of the g-value analysis in the later sections of this chapter.

In the calculation outlined above, it has been assumed that the unpaired electron remains entirely localized on the central transition group atom, but this will only be true, of course, for compounds in which the bonding is entirely of an ionic nature. In a large number of cases, however, a considerable amount of covalent bonding may also be present, and this is normally true for most organic and biochemical compounds. In such cases, the fact that the electron orbital moves out to embrace the ligand atoms to some extent, must also be considered, and the detailed theoretical calculations, associated with such transition group complexes, have come to be known under the general title of 'ligand field theory'. The basic steps that are followed when applying such a theory to the transition group complex, instead of the transition group atom by itself, follow the case of the single atom very closely, however, and can be summarized as follows.

The first step is to tabulate the energy level system of the central transition group atom on one side of an energy level diagram and, at the same time, tabulate the energy level system of the ligand atoms on the other side. The energy level diagram of the complex, can then be constructed by mixing the d-wave functions and orbitals of the central atom with those of the ligand atoms having the appropriate symmetry. This admixture results in a set of bonding orbitals, which have their main component from the ligand atoms, and a set of anti-bonding orbitals, which have their main components from the d-wave functions of the central metal ion. The effective ground state orbital of the previous calculation is then

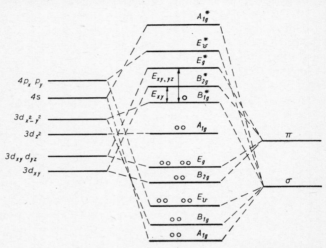

FIG. 6.6. Energy level systems of copper phthalocyanine complex

The energy levels of the Cu^{++} ion are shown on the left, while the π and σ levels of the ligand nitrogens are shown on the right. The levels appropriate to the complex are then found by admixing these, as shown in the centre, and the electrons belonging to the complex are then fed into these molecular orbitals, to leave one unpaired electron in the anti-bonding $B_{1g}*$ orbital

found by feeding the total number of electrons, belonging to both the $3d$ shell of the ion and the outer orbits of the ligand, into the molecular orbital pattern for the complex. These electrons are fed into these orbitals in pairs from the bottom upwards, and the lower bonding orbitals thus become filled with paired electrons and do not contribute to the magnetic properties of the complex. The remaining unpaired electron (or electrons) then resides in the anti-bonding orbitals, and the properties of the complex will be determined by the orbital containing the unpaired electron together with those that are immediately above it.

This method may be illustrated by the example of copper phthalocyanine, which is illustrated in Fig. 6.6. It will be seen that the orbitals corresponding to the Cu^{++} ion in its square configuration are shown on the left-hand side, while the two sets of ligand orbitals corresponding to the π and σ bonds on the nitrogen are given on the right-hand side. The energy levels of the molecular complex so formed are shown in the centre of the diagram; all the electrons in these outer orbitals are then filled in, as described, and it is seen that the last one enters the B_{1g}^* anti-bonding orbital, which lies below the B_{2g}^* and the E_g^* anti-bonding orbitals, as shown. All of the levels below the B_{1g}^* are filled with paired electrons and therefore do not contribute to the magnetic properties of the complex, which will be determined entirely by the presence of the unpaired electron in the B_{1g}^* orbital together with the admixture from higher orbitals.

The actual g-values to be associated with this particular complex are then calculated in exactly the same way as described earlier in this section for the ionically bound copper atom, but the relevant orbital levels and orbital level splittings are now those corresponding to the B_{1g}^*, B_{2g}^* and E_g^* levels of complex, rather than those of the d-orbitals of the copper atom itself.

In general it may be said that four effects can be expected, when any significant amount of covalent bonding takes place in such a complex, and these can be summarized as

(i) The divergence of the g-value from $2\cdot0$ is decreased, since there will be a reduction in the orbital contribution from the central metal ion.

(ii) The splitting of the hyperfine structure observed from the nucleus of the metal ion will also be decreased, since the interaction of the molecular orbital with the nucleus of the central ion will not be so strong as it was for the single atomic orbital.

(iii) There will most probably be an appearance of superhyperfine structure, arising from the magnetic moments of the ligand nuclei, since the molecular orbital now embraces these to some extent.

(iv) There will also be an increase in the spin lattice relaxation time, since this is determined to a large extent by the spin-orbit coupling, and the reduced orbital contribution from the atom will reduce the interaction and lengthen the relaxation time.

The way in which these effects can be observed, and used to study the covalent binding, is illustrated in detail in the next section, and the binding of the copper in copper ceruloplasmin is a good example. If single crystals of the compounds are available very considerable information on the amount of covalent bonding can often be obtained from a detailed analysis along the lines indicated above, and changes in both the g-value divergence and the hyperfine splitting, together with magnitudes of any superhyperfine structure, can be used to derive precise values for the parameters measuring the covalency present.

§6.4

PROTEINS CONTAINING COPPER

The way in which the actual magnitude of the g-values observed in an electron resonance spectra, together with the measured hyperfine splittings, can be used to give precise details on the nature of the chemical bonding, can probably be best illustrated by some examples taken from proteins containing copper ions of different valency. As was mentioned in the last chapter, the enzyme ceruloplasmin is one such protein, containing eight atoms of copper per molecule, and chemical analysis had suggested that there were probably approximately equal amounts of the cuprous and cupric copper present in this protein. Malmstrom[5] has, in fact, suggested that the cuprous copper is responsible for the binding of the substrate and that electron transfer from the substrate to the cupric copper is assisted by the Cu^+ ions acting via the π bonding of the aromatic ring of the substrate. Such a mechanism as this will of course imply fairly strong interaction between the two valency states of the copper atoms.

The observed E.S.R. spectrum from the ceruloplasmin[4] has a $g_\parallel$ of 2·209, and a $g_\perp$ of 2·056. There is also a well-resolved hyperfine splitting centred around the $g_\parallel$ direction, similar to that illustrated in Fig. 1.12(b), and this corresponds to a hyperfine splitting parameter, along the parallel direction, of 0·008 cm^{-1}. It will be seen immediately that these g-values have a departure from the free electron spin significantly less than that normally observed for cupric ions in a simple ionic bonding. This therefore suggests that there must be considerable covalency in the chemical bonding which will reduce the g-factor divergence from 2·0 as indicated in equation (5.1). In the same way, the actual hyperfine splitting observed in the parallel direction is much smaller than normal, as predicted by the second point made in the last section as characteristic of covalent binding in an atom. Apart from these general qualitative considerations, however, precise quantitative comparisons can also be made along the lines outlined in the last section. These have, in fact, been carried through by Vanngard and Aasa,[6] who assumed that the site of the copper atom had an axial field symmetry and that the absorption lines themselves would individually be of a Lorentzian shape. They were then able to predict the kind of overall pattern that would be obtained when an integration over the anisotropic g-value and hyperfine splitting had been made, and their theoretical analysis predicted values of $g_{\parallel} = 2 \cdot 214$, $g_{\perp} = 2 \cdot 048$, and $A = 0 \cdot 0083$ cm^{-1}, which values are seen to be in good agreement with those observed experimentally.

A very systematic analysis of the spectra observed from biological complexes containing copper has been made by Malmstrom and Vanngard,[7] who developed a technique of calculating two-dimensionless constants from the experimentally measured parameters of the spectra. These two constants could be used to characterize the nature of the bond to the metal atom, one being dependent on the shift in the g-value and the other being based on the hyperfine splitting. For ceruloplasmin the values of both these constants indicated a high degree of covalent bonding between the metal and the ligand. Thus, the value of the constant derived from a shift in g-value, which should be $1 \cdot 0$ for a purely ionic link and $0 \cdot 5$ for a purely covalent complex, was found to be $0 \cdot 52$ for the case of ceruloplasmin.

In a similar way, the constant determined from the hyperfine splitting suggested strong covalent bonding, as has already been seen qualitatively from the fact that the hyperfine splitting in the parallel direction is much lower than is obtained in normal ionic salts. As mentioned in the last section, further evidence for such covalent bonding can sometimes be obtained from the superhyperfine structure produced from the surrounding ligand nitrogen atoms, as has been shown in Fig. 5.3.

More recent investigations by Vanngard[8] on highly purified samples of ceruloplasmin show that, as well as the spectra with the small hyperfine splitting discussed above, there is also a signal due to copper atoms with a larger and thus more normal hyperfine splitting; in the preparations most mildly treated, the intensity of this additional signal corresponded to one copper atom per molecule. It would, therefore, appear that, of the four cupric ions in the ceruloplasmin molecule, one is bound in a different way from the others and the complete identification and location of these various copper atoms needs further elucidation. It is, nevertheless, very interesting to see how such detailed information can be obtained on the precise chemical binding of the copper atoms in these proteins by a straight forward analysis of the electron resonance spectra.

The investigation of electron resonance spectra can also often be of considerable help when the site of the copper, or other transition group metal atom, is one of very low symmetry, indicating some particular stereo-chemical strain in the complex. A good example of this is afforded by the work of Blumberg and his colleagues[9] on the E.S.R. spectra observed from stellacyanin, which is a blue glyco-protein containing one single divalent copper

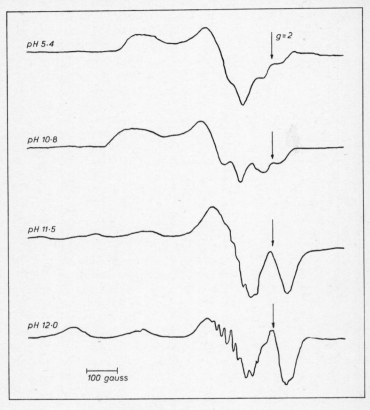

FIG. 6.7. E.S.R. spectra from copper atoms in stellacyanin

The *X*-band spectra of stellacyanin are shown for various *p*H values. Both the four hyperfine lines from the copper at the low field side, and the superhyperfine pattern from the nitrogens, can be clearly resolved at high *p*H values (From Peisach *et al.*[9])

atom per molecule. The E.S.R. spectra obtained from stellacyanin at different values of *p*H are shown in Fig. 6.7, and it can be seen that the hyperfine structure from the copper atoms, and also the superhyperfine structure from the surrounding nitrogens, becomes clearly resolved at the higher *p*H values.

The line corresponding to the $g_\perp$ direction has two major components associated with it, as well as the superhyperfine pattern. This implies that the electrostatic field acting on the copper atom does not have axial symmetry, and hence there is no longer an isotropic g-value, but different values for g_x and g_y. A detailed analysis of these curves gives for the g-values $g_z = 2\cdot30$, $g_y = 2\cdot06$ and $g_x = 2\cdot03$, and thus noticeable distortion of the molecular field from axial symmetry is present. These g-values have, moreover, been checked at higher frequencies to ensure that they are genuine g-value variants and not other forms of splitting in the spectra.

The appreciable rhombic character which must be present in the ligand field to explain this difference between g_x and g_y is also supported by the very intense optical absorption which is observed. Both the E.S.R. measurements and those on optical absorption indicate that the symmetry of the molecular field around the copper atom changes in two distinct steps as the pH value itself is altered. This might, in turn, be linked with the onset of denaturation in the protein: thus a strong nonaxial environment around the copper must imply a rigid matrix for the surrounding ligand atoms, since otherwise they would reorientate themselves to maintain the axial symmetry. The fact that this non-axial symmetry occurs at high pH values seems to suggest that noticeable changes may take place in this surrounding matrix, corresponding to denaturation of the tertiary structure of the protein at the high pH values. This particular line of study is thus of considerable interest in showing that onset of denaturation of proteins might best be studied in some cases from the changes in the axial symmetry of the electron resonance spectra.

§6.5

BIOCHEMICAL MOLECULES CONTAINING COBALT

There are several important biochemical compounds which contain cobalt as an essential constituent of the molecule, and these include in particular, vitamin B_{12}, and its related derivatives. The cobalt ion can exist in a variety of magnetic states when bound within an organic complex. The Co^{3+} ion has six $3d$ electrons, and is thus similar to the case of the ferrous ion discussed in detail in the next section. In any complex with strong covalent bonding the six electrons are forced into the lower three $3d$ orbitals, and thus all pair to give a diamagnetic molecule. Only in the weakest binding, such as with the fluoride atom, will the high spin state of $S = 2$ be obtained (as may be seen for the similar case of Fe^{++} in Table 6.1). Although this is paramagnetic, the zero-field splitting between the spin levels may be too large to allow the observation of electron resonance. This point is also discussed in detail in the next section, and the net result is that electron resonance spectra from Co^{3+} or Fe^{2+} ions are highly unlikely, unless experiments can be performed on single crystals at low temperatures with high microwave frequencies.

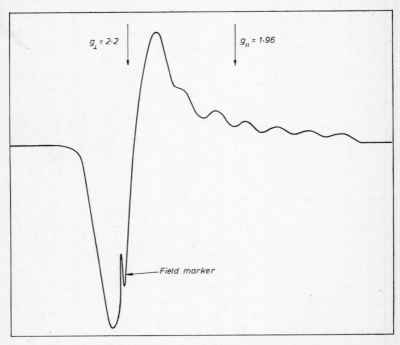

$g_\perp = 2.2$

$g_{||} = 1.96$

Field marker

FIG. 6.8. E.S.R. spectrum of cobalt in irradiated coenzyme B_{12}
The coenzyme B_{12} was irradiated at 0°C. Eight hyperfine lines centred on a $g_{||}$ of 1·96 can be seen, while the large low field absorption corresponds to a $g_\perp$ of 2·2 (From Hogenkamp *et al.*[12])

The Co^{2+} ion has seven $3d$ electrons, however, and is thus paramagnetic in both the strongly-bound and weakly-bound cases, the former corresponding to $S = \frac{1}{2}$ and the latter to $S = \frac{3}{2}$. In ordinary inorganic compounds the Co^{2+} is normally found in the high spin $S = \frac{3}{2}$ state, and there are quite a number of low-lying orbital levels. As a result, the g-values of the E.S.R. absorption are shifted very appreciably from $g = 2$ and the spectra can normally be observed only at low temperatures, because of the short spin-lattice relaxation times.[10, 11]

This brief summary of the magnetic properties of cobalt ions suggests that their presence in biochemical molecules will not be readily detectable by E.S.R. Early work showed that no E.S.R. signals could be detected directly from vitamin B_{12} and its derivatives, and it was concluded that the cobalt was probably present in the strongly-bound Co^{3+} diamagnetic state. On the other hand, Hogenkamp *et al.* have been able to obtain signals typical of Co^{2+} from samples of coenzyme B_{12} after irradiation. Such a signal is shown in Fig. 6.8, and can be positively identified as due to the cobalt from the eight-line hyperfine pattern due to the nuclear spin $I = \frac{1}{2}$ of the Co^{55}.

It would appear from this that an oxido-reduction takes place during the photolysis, and this example is a useful illustration of how E.S.R. can often follow these photochemical changes.

§6.6

STUDIES ON HAEMOGLOBIN AND RELATED DERIVATIVES

6.6.1 *The haem plane and its surroundings*

It was seen in various examples in the last chapter that there are a large number of important biochemical molecules which contain the haem, or porphyrin, plane as an essential part of their constitution. Thus catalase and peroxidase both contain this plane as a significant part of their structure, and it is also present in such enzymes as the cytochromes. This large flat molecular plane, which has already been shown diagrammatically in Fig. 5.8, is also one of the essential features of the structure of the haemoglobin molecule. In all these cases, a single iron atom is held in the centre of the haem plane, surrounded by a square of four nitrogen atoms, and the link to the protein part of the molecule itself is then made by a fifth coordination point on the iron atom itself.

In the case of haemoglobin, which is the molecule actually responsible for oxygen transport in the blood stream itself, there are four of these haem planes and associated iron atoms per molecule, whereas in the case of myoglobin, which is the similar molecule responsible for the storage and transport of oxygen within the muscles, there is only one haem plane and iron atom per molecule. Thus, to a first approximation, the haemoglobin can be considered as made up of four myoglobin components. In both cases, the sixth coordination point to the iron atom is taken by the oxygen, or carbon dioxide, and it is this particular bond which is intimately associated with the actual respiratory process. It is also possible for other chemical groups to be substituted on to this sixth co-ordination point, and, for example, in its normal stored state in typical buffer solutions, a water molecule is present in place of the oxygen or carbon dioxide, and this is then termed 'acid-met' haemoglobin or myoglobin.

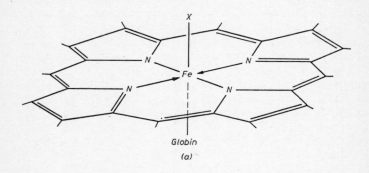

(a)

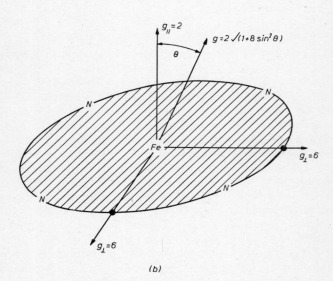

(b)

FIG. 6.9. Structural diagram of the centre of haemoglobin molecule

(a) The iron atom is shown at the centre of the haem, or porphyrin, plane, with the globin protein attached at the fifth coordination point. The oxygen, carbon dioxide, or other group X, is then attached at the sixth coordination point

(b) The g-value variation for the high spin haemoglobin derivative is shown, with a $g_\perp = 6 \cdot 0$ in all directions in the haem plane

The particular group and nature of the bonding at this point are, of course, crucial in the action of the haemoglobin. Thus, it is possible for fluoride ions to be substituted in place of the water molecule, and these can then also be readily removed. It is also possible to substitute a much more strongly bound group, such as cyanide, which of course is impossible to remove, so that it completely quenches the activity of the molecule. The location of the iron atom in the centre of this haem plane and its coordination with the protein at the fifth coordination point, and the other group at the sixth coordination point, are shown diagrammatically in Fig. 6.9. It will be appreciated that considerable interest arose as to how much information electron resonance could give on the nature of the binding between the iron atom and the crucial group on the sixth coordination point. As is often the case with research, it transpired that some of the most interesting information actually came from quite a different direction, and that electron resonance was not only able to provide details on the nature of the chemical binding, but also very significantly help in the elucidation of the structure of the molecule itself.

6.6.2 *High-spin and low-spin states of the iron atom*

Before discussing the electron resonance signals observed from haemoglobin and its derivatives, some consideration must be given to the actual states of the central iron atom and the energy levels associated with them. The iron atom can exist in either a ferrous or ferric state and, to a first approximation, either of these can be considered as bound ionically or covalently. There are thus four ways in which the outer electrons of the iron atom may be considered as occupying the $3d$ and $4s$ orbitals, and these are illustrated in Table 6.1. In this table each of the $3d$ and $4s$ orbitals is indicated as a circle, and each electron as an arrow, with its spin aligned up or down. The ferrous Fe^{++} ion has six electrons in its $3d$ shell to distribute amongst these orbitals, while the ferric Fe^{3+} has five.

If ionic bonding is present, there will be no sharing of electrons with the ligand atoms, and hence all of these orbitals will be available for the $3d$ electrons of the iron atom itself. Following the normal Hund's rule, these will therefore enter the orbitals to maximize the electron spin; hence, in the case of the ferric derivatives, all five electrons will enter five different $3d$ orbitals, with their spins aligned parallel, to give a total electron spin of $S = \frac{5}{2}$, whereas the sixth electron of the ferrous iron will have to enter an orbital which already contains one spin, as indicated in the table, giving a resultant total spin $S = 2$. If strong covalent bonding exists to the octahedron of ligand atoms on the six coordination points, there will be a sharing with the electrons of these ligand atoms and the $d^2s\,p^3$ orbital required for this will, therefore, take up the two higher $3d$ orbitals of the iron atom as well as the outer $4s$ orbitals, as indicated in the lower half of Table 6.1. In this case of covalent bonding, the six electrons of the ferrous iron are forced into the three remaining $3d$ orbitals, and thus all of them pair to give a total spin $S = 0$ and, thus, diamagnetic compounds. In the case of the ferric derivatives, the five electrons enter these three orbitals and, as a result, leave one unpaired spin with $S = \frac{1}{2}$.

TABLE 6.1

		Electron configuration			Total spin
		$3d$	$4s$	$4p$	
Ionic bonds	Fe^{++}	⊕①①①①	○○	○○○	2
	Fe^{+++}	①①①①①	○○	○○○	$\frac{5}{2}$
Octahedral covalent d^2sp^3 bonds	Fe^{++}	⊕⊕⊕ ○○	○○	○○○	0
	Fe^{+++}	⊕⊕① ○○	○○	○○○	$\frac{1}{2}$

The difference between the ionic and covalent bonds as represented in Table 6.1 can be also considered in terms of the energy level diagram of Fig. 6.2(c). It was seen here that the first effect of an octahedral symmetry on the $3d$ orbitals was to separate these into two groups, the lower one containing the three d_{xy}, d_{yz} and d_{zx} orbitals, while the upper contained the two $d_{x^2-y^2}$ and d_{z^2} orbitals. The actual way in which electrons are fed into these orbitals will depend crucially on the splitting between these two groups and the interactions between the electrons themselves. Thus, if the splitting between the two groups is rather small, as corresponds to the ionic case, the electrons will feed in, one at a time, to each orbital before they are paired. If, however, the splitting between the two groups of orbitals is large, as corresponds to the covalent case, more energy will be required to put the electrons into a higher orbital than to pair them in the lower group, and the pairing, with resultant smaller spin values, thus takes place.

It can be seen from this analysis that both the ferrous and ferric derivatives can thus be subdivided into 'high-spin' and 'low-spin' compounds, and these are effectively a measure of the strength of the binding between the iron atom and its surrounding ligands. We shall see later, however, that this classification is only a first approximation, although it serves as a very useful preliminary analysis of the binding of the iron atom.

The case of the low-spin derivatives may be considered first, since these are relatively simple. Thus the ferrous iron in the covalent bonding produces a zero resultant electron spin, and hence all the compounds containing covalently bound ferrous iron will be diamagnetic, and no electron resonance spectra will, therefore, be obtained from them. Ferric derivatives, on the other hand, will possess one unpaired spin per iron atom, and thus a single degenerate energy level. This will split on the application of a magnetic field in the ordinary way, to give g-values spread around the free spin value. The anisotropy of these g-values will reflect the surroundings of the iron atom and can be used to probe their symmetry, as discussed in later sections.

The case of the high-spin derivatives is, however, rather more complex, and both the ferrous and ferric cases require careful consideration. There is a very general theorem due to Kramers,[13] which states that, if an atom contains an odd number of electrons, the internal electrostatic fields can never remove the degeneracy of the levels completely and there will always be at least a two-fold degeneracy which can be removed only by the application of an external magnetic field. If, however, the paramagnetic atom contains an even number of unpaired electrons, it is possible for the internal molecular field to remove the degeneracy of all the levels completely, so that even those corresponding to $M_S = \pm\frac{1}{2}$ are split in the absence of the external magnetic field. Moreover, it is possible for this zero-field splitting to be larger than the energy corresponding to the microwave quantum used in the electron resonance spectrometer. If this is so, it will be impossible to observe electron resonance absorption from this system of energy levels, since the available quantum energy will never be able to induce transitions between the two lowest levels. This situation is, in fact, often found for the case of ionically bound ferrous derivatives, and hence very little information has been obtained by E.S.R. on such compounds.

This comment does not apply to the ferric derivatives, however, since these do possess an odd number of unpaired electrons, and therefore Kramer's theorem applies to them and each level will still be doubly degenerate in the absence of an externally applied magnetic field.

6.6.3 *Energy levels and anisotropy of the high-spin state*

It has already been seen that the five unpaired electrons associated with the high-spin ferric state align parallel to each other to give a total $S = \frac{5}{2}$, which may be represented vectorally as in Fig. 6.10. This total spin quantum number can itself now take up different orientations with respect to the axis of the internal molecular field, and $(2S+1)$, i.e. 6, orientations are then possible, as indicated in the figure, and characterized by quantum numbers $M_S = +\frac{5}{2}$, etc. According to Kramer's theorem, however, the energies of the $+\frac{5}{2}$ and $-\frac{5}{2}$ levels will be equal; in fact, these six orientations produce only three pairs of energy levels, as indicated in Fig. 6.10(b). The energy splitting between these three pairs of levels will now depend crucially on the magnitude and symmetry of the molecular field and, in most cases, the splitting is small compared with the microwave quanta being employed. A good example of this is the case of Mn^{++}, which is often observed as an impurity in biochemical studies, and the splitting between these three pairs of levels for such manganese salts, and the way in which they further split on the application of an external magnetic field, has already been considered in § 1.4.3 and illustrated in Fig. 1.7. It was seen there that, at any reasonable values of applied magnetic field, the resultant electron resonance spectra will consist of five groups of lines between these six different levels, and the zero field splitting produced by the molecular field will be only a small perturbation on the main splitting produced by the applied magnetic field.

In the case of haemoglobin and its related derivatives, however, the situation is entirely different. The internal molecular field now produces a very large splitting between the three pairs of energy levels, very much larger, in fact, than any normal microwave quanta. As a result, the microwave quanta are unable to induce any transitions between the higher levels and the ground state, and, at normal E.S.R. spectrometer frequencies, the only resonance to be observed will be that between the $M_s = \pm\frac{1}{2}$ levels of the lowest state itself. This situation is illustrated in Fig. 6.11. Moreover, the calculation of the effective g-values for this ground state will now be quite different from that of the normal ^6S ground state of manganese. Instead of the zero field splitting between these different spin states being considered as a small perturbation on the main interaction produced by the external magnetic field, the

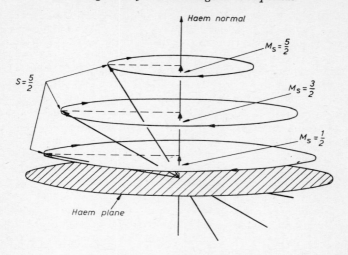

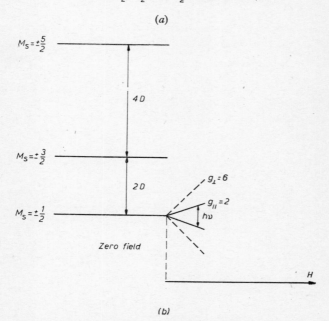

FIG. 6.10. Total spin orientation of ferric ion

The total spin of $S = \frac{5}{2}$ for the ionically bound ferric ion is shown in different orientations with respect to the molecular field axis in (a), and the energies of the different orientations are given in (b)

situation is completely reversed, and as a result, the axis of the internal molecular field is effectively the axis of quantization instead of that of the applied magnetic field. When the magnetic field is applied along the axis of the internal field, there is no competition between these, and the observed g-value will then correspond to that of the free spin, as expected for this singlet state in which there is no orbital contribution. On the other hand, there is significant admixture from higher spin levels into the ground state.

The result of the admixture of these higher spin levels is to change the effective g-value observed in directions away from the parallel axis, and in the case of the axial symmetry, an effective $g_\perp$ of 6·0 is obtained. The reason for this high effective $g_\perp$ can probably be appreciated qualitatively by a reference to Fig. 6.10. It is seen there that the particular orientation of the total vector of $S = \frac{5}{2}$, which corresponds to the ground state $M_S = \pm\frac{1}{2}$, is making a very wide angle with the molecular axis as it precesses. There will thus be a large component of magnetization in the directions corresponding to the haem plane (i.e. perpendicular to the axis), and hence a large g-value in these directions is to be expected. The more accurate derivation of this g-value follows from quantum-mechanical treatment of the energy levels concerned, and those familiar with this treatment may note that the secular determinant for the case of the magnetic field parallel to the molecular axis takes the form

$$\begin{array}{cc} \qquad\qquad |\tfrac{1}{2}\rangle & \qquad\qquad |-\tfrac{1}{2}\rangle \\ \begin{array}{c} \langle +\tfrac{1}{2}| \\ \\ \langle -\tfrac{1}{2}| \end{array} \left| \begin{array}{cc} \dfrac{D}{4}+\tfrac{1}{2}g\,\beta H_z - E & \quad 0 \\ \\ 0 & \dfrac{D}{4}-\tfrac{1}{2}g\,\beta H_z - E \end{array} \right| \end{array}$$

Whereas the secular determinant corresponding to the case of a magnetic field perpendicular to the molecular axis takes the form

$$\begin{array}{cc} \qquad\qquad |\tfrac{1}{2}\rangle & \quad |-\tfrac{1}{2}\rangle \\ \begin{array}{c} \langle +\tfrac{1}{2}| \\ \\ \langle -\tfrac{1}{2}| \end{array} \left| \begin{array}{cc} \dfrac{D}{4}-E & \tfrac{3}{2}g\,\beta H_x \\ \\ \tfrac{3}{2}g\,\beta H_x & \dfrac{D}{4}-E \end{array} \right| \end{array}$$

The solution of these two determinants then leads to expressions for the energy difference between the two levels, which become, respectively,

for H parallel to haem normal
$$\Delta E = g_\parallel{}^* \beta H_z \qquad \therefore g_\parallel{}^{\text{eff}} = g_\parallel{}^* \tag{6.3}$$

for H perpendicular to haem normal
$$\Delta E = 3g_\perp{}^* \beta H_x \qquad \therefore g_\perp{}^{\text{eff}} = 3g^* \tag{6.4}$$

In these expressions the g^*'s correspond to the true g-values, or those which will be observed if the zero field splitting is very small, whereas the g^{eff}'s correspond to those which are actually measured, and in the rest of this chapter these will be the g-values that are actually quoted.

The net result of this analysis is to show that, when a large zero field splitting between the spin levels is produced by the molecular field, the spectrum can be described by a fictitious spin of $S = \frac{1}{2}$, with $g_\parallel = 2$ and $g_\perp = 6$. There will, of course, be intermediate cases when electron resonance spectrometers operating at very high frequencies and short wavelengths are employed, and then the zero field splitting may be of the same order as the microwave quanta themselves. In such cases, a much more careful quantum mechanical treatment of the energy levels is required, with no approximations being employed, and the expression for the effective $g_\perp$ then becomes

$$g_\perp{}^{\text{eff}} = 3g_\perp{}^* \left[1 - 2\left(\frac{g_\perp{}^*\beta H}{2D}\right)^2 \right] \tag{6.5}$$

where $2D$ measures the splitting between the zero field spin doublets, as shown in Fig. 6.11. Hence as higher microwave frequencies are used, so the measured g-value in a perpendicular direction will be decreased from the value of 6·0, and it will be possible to use this shift of the $g_\perp{}^{\text{eff}}$ to estimate the actual splitting between the spin doublets themselves.

The actual way in which the energy levels diverge as the magnetic field is applied in this perpendicular direction is illustrated in Fig. 6.11, where the way in which the measured g-value will change from 6 to 2 as the field increases can be seen. It is also evident from this figure that, if sufficiently high microwave frequencies are available, it may be possible to observe transitions between the ground state doublet, and a component of the $M_S = \pm \frac{3}{2}$ doublet above. It should be noted, however, that better conditions for the observation of this direct transition are obtained when the applied magnetic field is closer to the direction of the haem plane normal. The more recent work on haemoglobin and its derivatives[14]

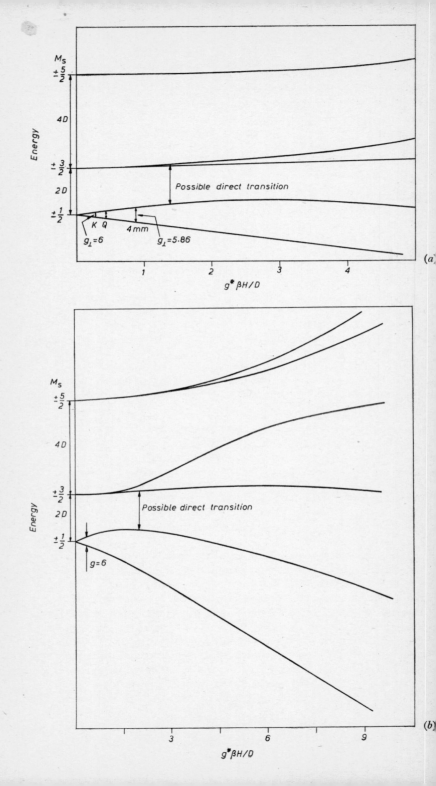

has, in fact, been concentrating on measurements at these high frequencies to determine the shift of the $g_\perp^{eff}$ away from 6·0 and also to obtain direct measurements of the $\pm\frac{3}{2}$ to $\pm\frac{1}{2}$ transitions.

In the early work,[15] however, measurements were made at microwave frequencies of either the X-band or Q-band region, in which, to a good approximation, the $g_\perp^{eff}$ was 6·0, and the anisotropy of the measured g-value for this lowest transition could therefore be represented as in Fig. 6.9(b). It can be seen from this figure that the g-value anistropy does, in fact, give a highly sensitive method of determining the orientation of the haem, or porphyrin, plane itself. In principle, all that is required in order to determine the orientation of the haem plane is to mount a crystal in the cavity resonator and then rotate the crystal in all possible directions until the g-value equal to 2·0 is obtained. It will then be known that the magnetic field is being applied along the direction normal to the haem plane; hence, the orientation of the plane itself is determined. In practice, however, it is very difficult to move a crystal in three directions at once, inside a cavity resonator. The crystals were therefore mounted in different crystallographic planes in turn, and the g-value variation for these different planes was then plotted. When these initial electron resonance investigations were made, the X-ray studies of Kendrew[16] and Perutz[17] had not been completed, and hence no information on the orientation of these planes was available. The orientation of these planes was, in fact, first determined by these electron resonance measurements, and this information was then handed to the X-ray crystallographers and used by them to assist in the complete analysis of the rest of the molecule.

6.6.4 *Haem-plane orientation as determined from high-spin g-values*

For accurate measurements of the orientation of molecular planes, it is of course essential that electron resonance measurements should be made on single crystals of the compounds concerned. In this connection, it is very fortunate that some bio-

---

FIG. 6.11. (*facing*) Splitting of high-spin Fe^{+++} levels on application of a magnetic field

The splitting of the spin doublets in zero magnetic field is shown at the left. The strength of the applied field is then given in terms of $g^*\beta H/D$, and the divergence of the energy levels is plotted for H perpendicular to the molecular field axis. It can be seen that, for low H, $g_\perp^{eff}$ is 6·0, but reduces from this as H increases. A possible direct transition between the $\pm\frac{1}{2}$ and the $\pm\frac{3}{2}$ levels, is shown by the vertical line.

(a) Large scale for small values of H
(b) Smaller scale for large values of H

chemical compounds can be grown as large single crystals, and haemoglobin and its various derivatives serve as a very good example of this. The initial X-ray studies of Perutz and Kendrew[16, 17] had shown that crystals of haemoglobin and myoglobin could be grown sufficiently large for X-ray studies to be possible, and by following up these techniques of crystal growing[18] it proved possible to produce well-formed crystals with edges some millimeters long in all three directions. It so happens that the myoglobin

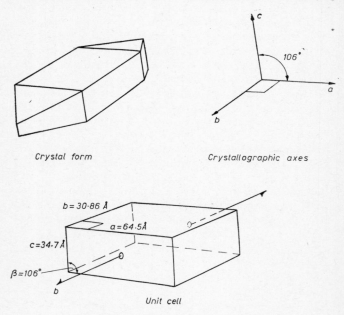

Crystal form   Crystallographic axes

Unit cell

Fig. 6.12. Type A myoglobin crystals

The well-formed faces, corresponding to the *ab* plane can be clearly seen, together with the *c* axis

from sperm whale is the easiest to grow as such large well-formed crystals, and the method of converting the initial whale muscle into single crystals, was initially developed by Kendrew and Parish[19] at the Cavendish Laboratory. The final stage of crystallization can take place either in the presence of ammonium sulphate buffer solutions or in the presence of phosphate buffer solutions at a pH of 7. In the former case, the crystals grow in the form known as type A, with the *ab* crystallographic face well developed, and possess diamond-shaped facets at each end, while the *c* axis is well defined more or less at right angles to this, as shown in Fig. 6.12.

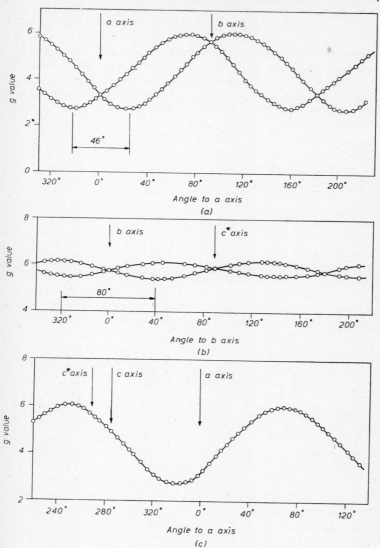

Fig. 6.13. *g*-value variations for type A myoglobin

The *g*-value variations for the acid-met type A myoglobin are shown for the crystallographic planes (*a*) *ab*, (*b*) *bc* and (*c*) *ac*. In the first two cases two curves are obtained from the two molecules per unit cell, whereas these project equally on the *ac* plane to give a single resultant curve

In the phosphate solutions, however, the crystals grow in quite a different form, known as type B, which is discussed in more detail later.

Although each myoglobin molecule contains only one iron atom and one haem plane per molecule, each unit cell of the crystal structure of the type A contains two molecules of myoglobin. There will thus be two differently orientated haem planes in such a crystal, and hence two sets of g-value variation are to be expected in the electron resonance results. In order to obtain a complete three dimensional map of these g-value variations, the type A crystal is first mounted in a H_{111} cylindrical cavity with the *ab* crystallographic plane horizontal on the bottom surface. The external magnetic field can then be rotated conveniently in all directions around the *ab* plane, and thus the g-value variation in the plane can be measured directly. The results obtained for the *ab* plane of a type A sperm whale myoglobin crystal are shown in Fig. 6.13(*a*). The two g-value variations, corresponding to the two molecules per unit cell, can be seen and it is evident that both of these reach a maximum value of $g = 6$, since any crystallographic plane must cut a haem plane in one direction at least, however it is orientated. The minimum g-value observed does not fall to 2·0 however, since neither of the normals to the haem plane, corresponding to the $g_{\parallel}$ direction, actually lies in the *ab* crystallographic plane itself. In fact the minimum observed g-value of 2·64, which occurs at an angle of 23° to the *a* axis, can be used to determine the angle between this minimum g-value direction and the haem normal. Thus it can be substituted into the equation

$$g_\theta{}^2 = 4(1 + 8 \sin^2 \theta) \tag{6.6}$$

to give a value for the angle θ between the haem normal and this direction, of $17\frac{1}{2}°$. It therefore follows that this single measurement of the minimum g-value in the *ab* plane and its orientation are sufficient to locate the direction of the normal to the haem plane itself.

Measurements are always carried out in other crystallographic planes, however, in order to cross-check the calculation, and the results obtained in the *bc* and *ac* planes are also shown in Fig. 6.13. It can be seen that the g-value variation in the *bc* plane stays very close to the maximum value of 6·0 in all directions, and this is to be expected since the two haem planes themselves have orientations quite close to that of the *bc* plane. Quite a wide g-value variation is again obtained in the *ac* plane, however, since this, like the *ab* plane, approaches quite close to the direction of the haem normals. Since this is a plane of symmetry, on which the two haem planes

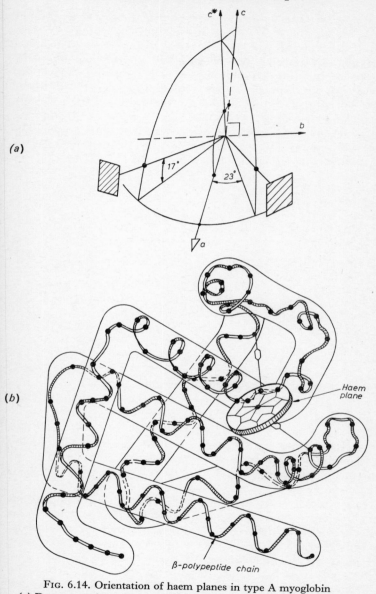

Fig. 6.14. Orientation of haem planes in type A myoglobin
(a) Drawn as a three-dimensional perspective plot, showing the angles of the normals to the *ab* plane and *c* axis
(b) In relation to the whole myoglobin molecule as determined from X-ray crystallography.[20] The haem plane is drawn in heavy outline

always project equally, only one *g*-value variation will be obtained
in this plane. Quantitative analyses of the *g*-values observed in
both of these planes, can again be used, as for the case of the *ab*
plane, to determine the orientation of the two haem planes, and
thus a complete cross check can be obtained. The orientation of the
two haem planes, as determined in this way, is shown in Fig. 6.14(*a*)
in the form of a three-dimensional perspective plot, with the
planes themselves indicated by the shaded squares. As mentioned
earlier, this determination by electron resonance of the orientation
of these haem planes, was obtained before the detailed X-ray
measurements on the myoglobin were available.[20] They were, in

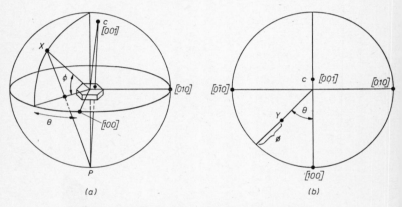

(a)   (b)

FIG. 6.15. Stereographic plot of *g*-value variation
(*a*) Principle of stereographic projection
(*b*) Stereogram so produced for Type A crystal axes

fact, used as an important piece of information in this final analysis,
and the complete structure of the myoglobin molecule, deduced
from this X-ray work, is shown in Fig. 6.14(*b*), with the haem
plane in heavy outline.

These results, which depend essentially on a three-dimensional
plot of *g*-value variation in order to locate the direction of the haem
plane normals, can in fact be better represented by the stereographic method of projection normally employed by X-ray
crystallographers. This method represents directions in three-
dimensional space by points on a two-dimensional plane. The
principle of the stereogram is illustrated in Fig. 6.15. A type-A
myoglobin crystal is taken there as a specific example and the
stereogram is constructed around it. In this method of projection,
a sphere can be imagined drawn around the crystal, and each

direction of interest can then by projected from the origin at the crystal until it intersects the spherical surface. These points of intersection on the spherical surface are then projected down on to the diametral plane of the sphere by joining them to the bottom pole of the sphere labelled P in the diagram. This direction in three-dimensional space is then denoted unequivocally by the single point on a diametral plane, labelled Y on Fig. 6.15(b).

It can also be seen from this diagram that the *ab* or (001) face of the myoglobin crystal, which is shown in a horizontal plane, will have both the *a* [100] and *b* [100] axes projecting out to cut the spherical surface on the equatorial circle, and the points corresponding to these two directions will therefore be on the circumference of the stereogram as labelled in the figure. On the other hand, since the type A myoglobin crystal is monoclinic, the *c* [100] axis is not normal to the *ab* plane, but projects away from this normal and intersects the spherical surface behind the upper pole; hence, its corresponding point on the stereogram is displaced from the centre as shown. It is also clear from this figure that the actual orientation of the general direction OX, represented by Y on the stereogram, is defined by the two angles θ and ϕ and that the value of both of these angles can also be deduced directly from the point Y on the stereogram itself.

It follows that the actual direction of a haem normal will correspond to a single point on this stereogram, since there is only one direction for a given haem plane in which the g-value falls to 2·0. On the other hand, all the directions which correspond to a g-value of 3·0 will lie on the surface of a cone which makes 23° to the haem normal, and these will therefore form a closed circle around the $g = 2·0$ on the stereogram. Similarly all the directions corresponding to $g = 4$ will lie on a cone making an angle of 38° to the haem normal; and those directions corresponding to $g = 6·0$, and thus to directions in the haem plane itself, will all be at right angles to the haem normal and will therefore lie on a great circle of the sphere and their projection can readily be found from a standard stereographic net.

Such a stereogram, on which the locus of constant g-values are plotted, as shown in Fig. 6.16 for the type-A myoglobin crystal, will give a complete summary of the g-value variation in all directions in three-dimensional space. In the actual experiments on the myoglobin and haemoglobin crystals themselves, the g-values, as determined from the measurements in the different crystallographic planes, are first plotted on such a stereogram and the points corresponding to the $g = 2·0$ can then be determined from these. The most direct method of doing this, is to plot out the points corresponding to a *g*-value of 6·0 as obtained from the different planes, and then these will fall on the projection of a great circle and the position of its normal can be found directly by the use of a standard stereographic net. However, as will be seen later, there are certain rhombic components which cause variations in this *g*-value of 6·0 and the most precise direction for the haem normal can be obtained by plotting successively smaller circles, corres-

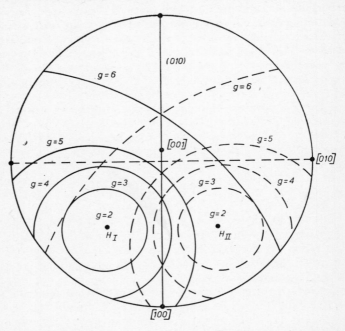

FIG 6.16. Complete stereogram for *g*-value variation in type A crystals
The directions of the haem normals correspond to the two points for which *g* = 2·0. They are labelled H_I and H_{II}

ponding to successively smaller *g*-values, until the single point *g* = 2·0 is located. As explained earlier, it is also possible to calculate these haem normal orientations without the use of the stereographic net, simply by substitution into equation (6.6). One of the most useful features of the stereograms, however, is that they enable all the E.S.R. information on these crystals to be presented in a straightforward way, which can be readily interpreted precisely by other workers in the field.

Following this work on the type-A myoglobin crystals obtained from the sperm whale, successive series of measurements were also made on other types of myoglobin crystals obtained from different species.[21] The orientation of the haem planes in these different crystals was determined in exactly the same way as for the type A crystal, and the results can be readily summarized both in the form of three-dimensional perspective diagrams and by the stereograms just described. The results for the type-B and the type-F crystals can be compared in this connection. Both of these crystals are of an orthorhombic form, with a space group P $2_12_12_1$ and possess four molecules per unit cell. Four different g-value variations are, therefore, to be expected in these crystals, and an example of the actual g-value plot, obtained when the magnetic field is rotated

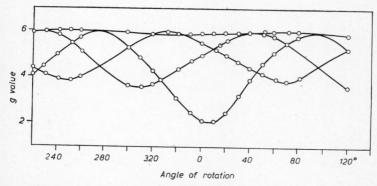

Fig. 6.17. g-value plot for type F crystal in plane containing haem normal
One g-value variation is seen to extend over the whole range from 2·0 to 6·0, while another stays close to 6·0 since its haem normal is at 84° to the plane of rotation

about a type F crystal in a plane containing one haem normal, is shown in Fig. 6.17. It can be seen that the g-value variation of this haem group extends over the complete range from 2·0 to 6·0, while another stays very close to the 6·0 value in all directions, since its haem normal is at 84° to the plane of rotation. The high symmetry of these two types of crystal can be seen in the two stereograms, which summarize the g-value variations, and which are shown in Figs. 6.18 and 6.19 together with the three-dimensional perspective diagrams for both cases. Since these two crystals are both orthorhombic, the three crystallographic axes will all be mutually perpendicular in this case, and the essential difference between the type-B and type-F crystals is really only in the different angles which their haem normals make to the *ab* plane and the *c* axis. This can be seen very clearly in the three-dimensional perspective

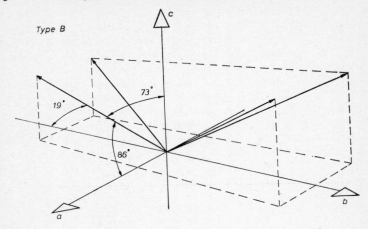

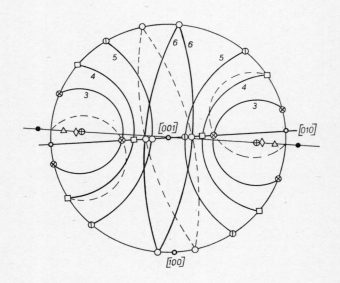

FIG. 6.18. Haem plane orientation for type B crystal

A three-dimensional perspective plot and a stereogram of complete *g*-value variations. The circles corresponding to constant *g* values of 6, 5, 4, and 3 are drawn as full lines for one pair of haem normals. The broken lines are the circles for *g* values of 6 and 3 for the other pair. Centres of circles: ●, $g = 5$; △, $g = 4$; ◇, $g = 3$. Direction of haem axis giving $g = 2$ is indicated by ⊕.

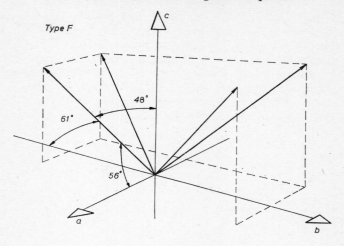

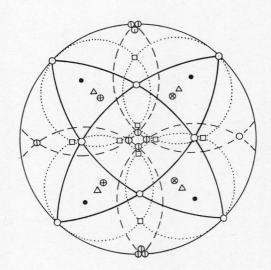

FIG. 6.19. Haem plane orientation for type F crystal

A three-dimensional perspective plot and a stereogram of complete *g*-value variations. The circles corresponding to values of 6 (———), 5 (– – –), and 4 (.......) are plotted for all four haem groups. The centres of these circles are also indicatad and are seen to lie on two straight lines intersecting at the origin. Centres of circles and direction of haem axis giving *g* = 2 are indicated by the same notation as in Fig. 6.18

plot, but it is also evident that much more detailed information is actually presented in the two stereograms beside them.

Following this initial work on the myoglobin crystals, electron resonance measurements were made on crystals of haemoglobin itself.[15] As mentioned earlier, haemoglobin contains four haem planes and iron atoms per molecule; and in the crystals of horse haemoglobin employed, there is only one molecule per unit cell. It had been assumed, before the electron resonance measurements

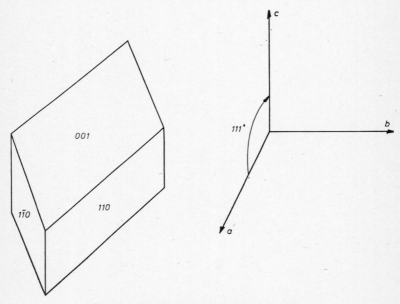

FIG. 6.20. Crystal of horse haemoglobin
It can be seen that the (001) and (110) faces are well developed

were carried out, that the four haem planes in the haemoglobin molecule were probably parallel to one another, but the initial electron resonance measurements indicated immediately that this was not the case. A typical crystal of horse haemoglobin is shown in Fig. 6.20, and it is clear that both the (110) and the (001) faces are well developed and readily identifiable. The g-value variations observed from the acid met derivative in these two planes, are shown in Fig. 6.21. It is quite clear that there are four separate g-value variations in both of these planes, and these correspond to four differently orientated haem planes per molecule. In the (100) plane, two of these g-value variations reach their absolute minimum of 2·0 and hence immediately define the actual directions of the haem normals of the two planes with which they are associated. The orientation of the other two planes can be found from the kind of quantitative analysis previously outlined for the myoglobin crystals or by the use of the stereographic plot, as described above. The results can then be summarized as a three-dimensional per-

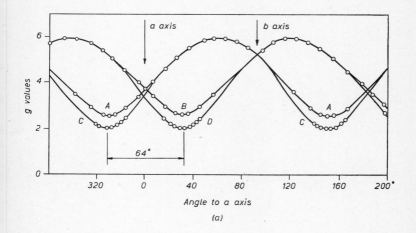

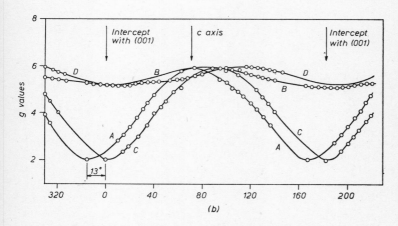

Fig. 6.21. g-value variation for haemoglobin crystal

(*a*) Variations in g-value in the (001) plane. It can be seen that two of the curves reach 2·0 and hence the orientations of their haem normals are immediately located

(*b*) Variation in the (110) plane. Another haem plane normal (labelled A) is also immediately located from its $g = 2·0$ in this plane

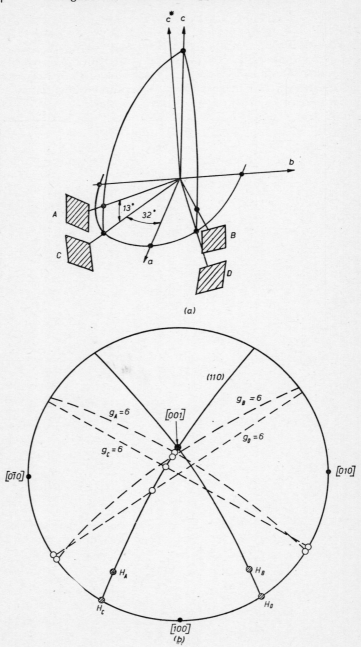

(a)

(b)

spective drawing or as a stereogram, as shown in Fig. 6.22(*a*) and (*b*).

Again it is clear that the electron resonance measurements have enabled the orientations of these four haem planes to be established unequivocally, and this information was obtained some time before the detailed X-ray analysis of these crystals was available. This particular work on the orientation of the haem planes in the myoglobin and haemoglobin derivatives is a very good example of how two different techniques, such as electron resonance and X-ray crystallography, can combine to give information of vital importance to each other.

It will be appreciated that the structural information, and the haem plane orientations, have been obtained in these cases by using the g-value variation purely as a probe of the symmetry around the axis through the haem plane. No detailed calculations on the actual magnitude of the g-values was necessary, and the arguments were really based entirely on those of pure symmetry, and the fact that the g-value parallel to the axis was so noticeably different from that in the perpendicular direction. In the next section it will be seen that this kind of analysis can be taken one stage further, however. Thus, the actual magnitudes of the g-values along their different directions can be used to determine the energy splittings within the groups of the electron orbitals of the central iron atom. This in turn can then give further structural information on the surroundings of the iron atom. This second stage is a very good example of the additional information that can often be obtained when a slightly more sophisticated analysis of the electron resonance results is undertaken.

6.6.5 *Structural information from low-spin g-values*

It has already been pointed out that the haemoglobin derivatives, which are strongly bound to the central iron atom, will possess only one unpaired spin in their $3d$ orbitals and hence have a resultant $S = \frac{1}{2}$. The g-values corresponding to the transitions $M_S = \pm\frac{1}{2}$ in this ground state will then be spread across the free spin value and a summary of those observed in the case of haemoglobin azide is shown in Fig. 6.23(*a*). It is seen that there is still a principal g-value close to the haem normal of 2·80, but that there is now no longer axial symmetry. Thus the g-values around the haem plane itself vary significantly from a minimum of 1·72 to a maximum of

---

FIG. 6.22. (*facing*) Haem plane orientation in haemoglobin

(*a*) As a three-dimensional perspective plot
(*b*) As a stereogram of g-value variations; the four haem normals are represented by H_A, H_B, H_C, H_D

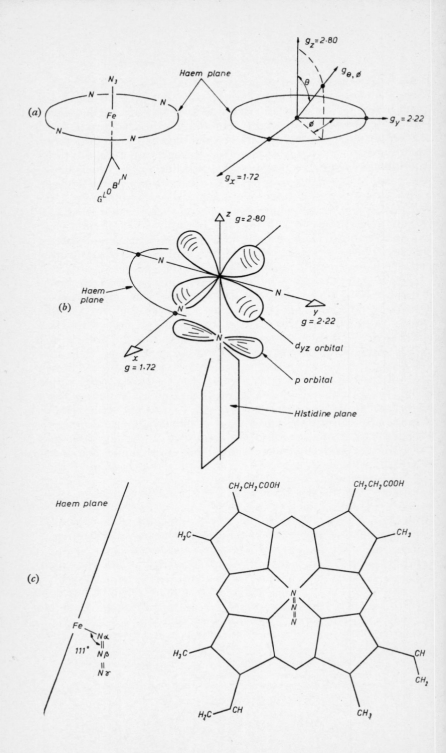

2·22. Three directions of principal g-values, g_x, g_y, g_z, can thus be defined and it must follow that the variation in g-value in the haem plane reflects some structural features above or below the plane, which are producing the anisotropic effect in the plane itself. The immediate structural features which might affect the orbitals of the iron atom are, on the one hand, the nitrogen atom on the fifth coordination point belonging to the histidine plane, which links the haem plane to the polypeptide chains of the protein. On the other hand, there is a nitrogen atom on the sixth coordination point which is one of the three nitrogen atoms of the N_3 azide group. It would appear, therefore, that some feature of either the nitrogen of the histidine plane below, or of the azide group above, is interacting with the orbitals of the iron atom to produce the anisotropy in the g-value around the haem plane itself.

The nature of this interaction can probably be best understood by referring back to Fig. 6.5. It was pointed out in §6.2 and discussed through the comments on Figs. 6.3 to 6.5, that the orbitals on the surrounding ligand atoms will shift the energy levels of the d-orbitals of the iron atom quite significantly. In particular, the p-orbital on the nitrogen at the fifth or sixth coordination point can interact appreciably with either the d_{zx} or the d_{yz} orbital, and force this orbital to a higher energy state. Fig. 6.5 makes it quite clear that if the p-orbital of the ligand nitrogen is along the x-axis, then the d_{zx} orbital of the iron atom will be raised to the higher level, whereas if it is along the y axis the d_{yz} orbital will be so raised. A straightforward molecular orbital treatment of the g-values observed in the azide derivative shows that these imply that the d-orbital corresponding to the direction of highest g-value in the haem plane must be lying highest in the t_{2g} group. Thus if the maximum g-value of 2·22 in the haem plane is associated with the y-direction, the molecular orbital analysis implies that the d_{yz} orbit will be lying highest in the t_{2g} group, and therefore that the p-orbital of the nitrogen of the histidine plane will be parallel to the y-axis. Since the histidine plane itself is at right angles to the p-orbital of its own nitrogen, it must therefore follow that the

---

FIG. 6.23. (*facing*) g-value variation for myoglobin azide

(a) Principal g-values as referred to the haem plane. There is no longer axial symmetry and the g-value varies from 1·72 to 2·22 around the plane itself

(b) Interaction of p orbit of histidine nitrogen atom with d_{yz} orbital on the iron atom. The high g-value along the Oy axis therefore implies that the histidine plane itself is parallel to Ox

(c) Orientation of azide group with respect to the haem plane (From Kendrew *et al.*[22])

plane of the histidine ring is along the x-axis as shown in Fig. 6.23(b).

So far, however, only the effect of the p-orbital of the nitrogen on the fifth coordination point has been considered, whereas there may well also be an interaction with the p-orbital of the nitrogen on the sixth coordination point, belonging to the azide group. It was initially assumed that the three nitrogens of this azide group were positioned along a projection of the haem normal and thus would not cause any asymmetry in the g-value around the plane. Recent X-ray work[22] has shown, however, that they are orientated along a line which makes an angle of $111°$ with the haem normal, as shown in Fig. 6.23(c), and projects down on to the haem plane as indicated. In this case, a significant electron density might well be expected in the p lobes of the nitrogen of the sixth coordination point, at right angles to the line of the azide group, and hence its orientation would also produce an asymmetry in the g-values in the haem plane, in the same way as discussed for the histidine plane above. It is found experimentally that the direction of the minimum g-value in the haem plane lies approximately between the projections of the histidine plane and the azide group, suggesting that both of these structural features are affecting the observed g-value variation. Measurements are in hand on the detailed g-value variation in cyanide and other covalently bound derivatives of myoglobin and haemoglobin, so that the contributions from the histidine plane and other groups on the sixth coordination point can be clearly differentiated.

It is evident from these present results, however, that a determination of the three-dimensional g-value variation in these covalently bound derivatives with low spin, together with a small amount of molecular orbital theory, does allow significantly more information to be obtained on both the structure of the haemoglobin molecule and the binding of the central iron atom.

Although attention has been focused on single-crystal studies throughout the last two sections, it should also be mentioned that the characteristic g-values of the high-spin and low-spin states can often be used as a means of identification, even in polycrystalline or liquid samples. This fact is well illustrated by the work of Morita and Mason[23] on the state of the iron atom in various peroxidases. They studied frozen solutions of these between $80°$ and $180°K$ and showed that absorptions at $g = 6·0$ and $2·0$ were present in all the samples. The temperature dependence of the absorptions showed that there was some thermal equilibrium between the low-spin and high-spin states and that the ratio of high-spin to low-spin state was also dependent on the pH value.

Moreover, conversion to the fluoride derivative produced much narrower signals at $g = 6 \cdot 0$, suggesting a complete removal of the low-spin state; whereas the covalently bound derivatives such as the cyanides, azides and hydroxides gave much stronger signals close to $g = 2 \cdot 0$.

These experiments indicate quite clearly that E.S.R. measurements can be applied directly to haem-containing proteins in conditions much closer to those *in vivo* as well as to the more specific single-crystal studies.

6.6.6 *The study of line widths and hyperfine splittings*

The widths of the electron resonance absorption lines, observed from the single crystals of myoglobin and haemoglobin derivatives, are much larger than those that would be expected from normal broadening processes. Thus the iron atoms are present in extremely small concentration, and the dipolar interaction between them would not produce broadenings of more than a few gauss at the most. The observed line-widths are also found to be independent of temperature from $4°$ to $90°K$ and hence there is no major contribution from the spin lattice relaxation effects. Instead of widths of the order of a few gauss however, line-widths of several hundred gauss are observed and these vary quite rapidly with orientation. This rapid variation with angle, and the fact that the line narrows to a significantly smaller width along the directions of the extreme g-values, gave the first clues to the broadening mechanism that was probably present. It will be appreciated that the resonance field position is very sensitive to the angle which the applied magnetic field makes with the haem normal, since there is such a rapid and large variation of g-value with angle in these compounds, and particularly so for the case of the acid-met derivatives. If, therefore, for some reason or other, there was a slight variation in the axial direction of haem normals from molecule to molecule, a major spread of the field positions required for resonance would be produced. This would, moreover, have just the kind of angular variation that was observed for the line-widths of the acid met derivatives. Detailed calculations were therefore undertaken to see if this randomization of the haem plane normals within the crystal might be able to account, both qualitatively and quantitatively, for the observed line-widths and their angular variation. In the case of the acid-met derivative, only one variable need be considered in such a theory, and this is the mean deviation in angle, associated with such a statistical variation of the orientation of the haem normals. This spread in orientation will produce a corresponding spread in g-values from molecule to molecule, and hence smear out the overall line-width. The actual expression for the line-width produced by such an effect can be written down quite explicitly,[24] as in the equation below

$$\Delta H = \frac{hv}{\beta}\frac{(g_\perp{}^2 - g_\parallel{}^2)}{2g_\theta{}^3}\sin 2\theta \sin 2\Delta\theta + \text{const.} \qquad (6.7)$$

where θ is the angle between the applied magnetic field and the haem normal, and $\Delta\theta$ is the standard deviation in the random

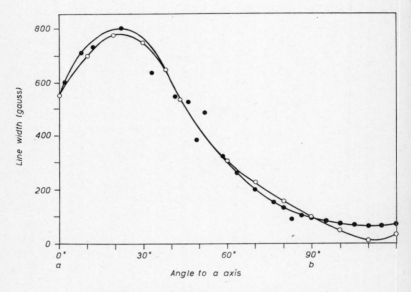

FIG. 6.24. Line-width variation in acid-met myoglobin

The experimentally measured values of the line widths in the *ab* plane are plotted as black circles while the open circles are calculated from equation (6·7)

orientation of the haem group directions. The first term in the above equation thus represents the contribution to the line-width from the spread in *g*-values associated with this random orientation, while the second term is a constant which includes the other residual broadening mechanisms. The fact that this expression does account for the variation in the line-widths to a very good approximation is shown in Fig. 6.24, where a comparison is made between the theoretical variation in line-width predicted by equation (6.7) and the experimentally measured values. The points represented by the open circles correspond to the theoretical variation given by equation (6.7), assuming a value for $2\Delta\theta$ equal to 0·055 radians, or 3·3°. It would appear, therefore, that a random orientation of only 1·6° in the direction of the haem normals is quite sufficient to account for the line broadening observed in the acid-met myoglobin.

Although the main features of the line-width variation of these haemoglobin and myoglobin absorptions can be explained in this way, more recent measurements have made it clear that there must also be some other mechanisms present which are not yet fully understood. Thus a systematic study of the line-width variation with angle at 4 mm wavelengths[14] shows that, as well as the variation which is produced by the g-value spread explained above, there is also an additional component which varies directly with the frequency of the microwaves used in the measurement. All normal mechanisms of line broadening are frequency independent, and hence these results must reflect some new type of broadening mechanism in these haemoglobin derivatives, which has not been encountered before.

One possible explanation is that there is not only a random variation in the haem plane orientation, but also a possible variation in the value of the zero field splitting parameter, D, brought about by variation in the distance of the water molecule from the iron atom in the acid-met derivative. Such a variation in this zero field splitting parameter, D, would in its turn produce a randomization in the effective g-value, as given by equation (6.5). Moreover, it is clear from equation (6.5) that the effective g-value depends on the actual ratio of the splitting parameter, D, to the frequency of the applied microwaves and hence to the value of the magnetic field strength required for resonance. This type of variation would therefore be one of the few mechanisms that would produce an additional broadening effect linearly dependent on the microwave frequency used for measurement. Much more work will have to be carried out, however, before all the problems associated with the line-widths of these compounds have been elucidated. Further details will not be considered here, but the results already quoted should be sufficient to show that systematic studies of E.S.R. line-widths in such derivatives can often give additional information on the basic interactions present in such molecules.

This additional line-width broadening, brought about by the random orientation of the haem normals, limits the resolution in the spectra and prevents the observation of any superhyperfine structure from the surrounding nitrogen atoms. It is possible that measurements on the crystals at room temperature might allow this, since a motional narrowing might then occur and the line-width be reduced accordingly. The large water content of the crystals has prevented such observations to date, however, although attempts to do this will undoubtedly be made in the near future. In the meantime, however, one hyperfine splitting from an adjacent ligand atom has been observed, and this is in the myoglobin fluoride derivative.[25]

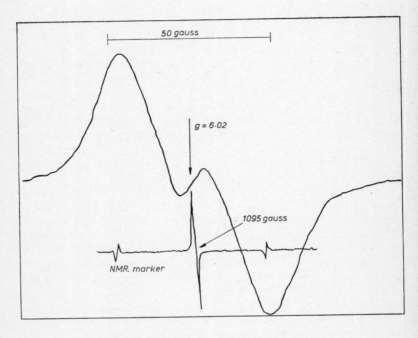

FIG. 6.25. Hyperfine splitting in myoglobin fluoride
The doublet splitting of about 50 gauss is due to the $I = \frac{1}{2}$ of the F^{19} nucleus on the sixth coordination point. (From Kotani and Morimoto [25])

The myoglobin fluoride is ionically bound, and hence corresponds to one of the high-spin derivatives like acid-met myoglobin. The g-value variation and spectra are thus very similar to that observed for the met-myoglobin. Kotani and Morimoto[25] were able to resolve a doublet splitting in this absorption, and this doublet can be attributed to a hyperfine coupling with the F^{19} nucleus, which has a spin $I = \frac{1}{2}$. A typical absorption line is shown in Fig. 6.25. The biochemical interest of this particular result is that it shows that there is considerable overlap between the un-paired electron on the iron atom and the ligand of the sixth co-ordination point, even in the case of the ionically bound high-spin derivatives. Further studies on hyperfine splittings observed from other groups on this coordination point should help to give much more precise details on the molecular orbitals associated with the central iron atom.

6.6.7 *Determination of the zero-field splitting parameters*

It has already been seen that one of the characteristic features of the high-spin myoglobin and haemoglobin derivatives, is the very large splitting produced between the energy levels of the spin components by the internal molecular field. It is this splitting which is responsible for the observed $g_\perp^{\text{eff}}$ of 6·0, as already shown in Fig. 6.11. The actual magnitude of this splitting and the mechanism which produces it are obviously closely related to the bonding of the iron atom itself. Several attempts have been made and are being made, therefore, to determine this splitting as accurately as possible in as many different derivatives as are available, so that a consistent picture of the energy level structure of this central part of the molecule can be derived.

There are, in essence, two ways in which the splitting parameter, D, can be measured. On the one hand, it can be determined directly if it is possible to induce a transition between one of the two levels in the ground state and one of the levels in the upper $\pm\frac{3}{2}$ compo-nent, as indicated by the vertical arrows of Fig. 6.11. As already explained, the magnitude of the microwave quantum at normal electron resonance wavelengths of 3 cm, and even 8 mm, is not sufficient to induce such a transition, and this is why the shorter wavelength 4-mm and 2-mm resonance spectrometers are being developed. The availability of such wavelengths, together with the high magnetic fields produced by superconducting magnets, should make it possible to observe direct transitions between the $\pm\frac{1}{2}$ and $\pm\frac{3}{2}$ spin levels, and hence direct determination of the value of the splitting parameter D will then be available. Prelimi-nary measurements reporting such observations, have, in fact, just been obtained.[26]

It is also possible, however, to obtain quite an accurate estimate of the splitting parameter, D, from the variation of $g_\perp^{\text{eff}}$ as measure-ments are made with ever higher microwave frequencies. Thus at low microwave frequencies, and hence at small quantum energies compared with D, the $g_\perp^{\text{eff}}$ will be 6·0; but, as the frequency of measurement is increased, so this $g_\perp^{\text{eff}}$ begins to be reduced from 6·0 as described by equation (6.5). A reference either to this

s

equation or to Fig. 6.11 shows that, at microwave frequencies of 72 000 Mc/s (i.e. wavelengths of 4 mm) or above, there will be significant shifts of the $g_\perp^{\text{eff}}$ from 6·0; hence these will enable the value of D to be estimated fairly accurately.

This analysis is somewhat more complex than might appear at first sight from equation (6.5), since the g-value variation in the haem plane is not quite isotropic. There is a rhombic component which can be represented by a splitting parameter E, as well as the axial splitting component D. The effect of this rhombic component is to produce anisotropy in $g_\perp^{\text{eff}}$, and hence g_x and g_y now have slightly different values for the acid-met as well as the azide derivatives. Various sets of measurements have been made on the different ionically bound myoglobin and haemoglobin derivatives to obtain values of the parameters D and E by this method.

Eisenberger and Pershan[27] were the first to deduce values of D from the divergence of $g_\perp^{\text{eff}}$ from 6·0. They undertook careful measurements at both 13 000 and 35 000 Mc/s. but did not make any allowance for the rhombic term E in their analysis. Their measurements of $g_\perp^{\text{eff}}$ equal to 5·950 and 5·930 at 13 000 and 35 000 Mc/s respectively then led to an estimate for $2D$ of 8·76 cm^{-1}. This work was then followed by the measurements of Kotani and Morimoto,[25] who measured the anisotropy of $g_\perp^{\text{eff}}$ in the haem plane, as well as its divergence from 6·0. They obtained estimates for D of 10·0 cm^{-1} for the acid-met myoglobin and 6·5 cm^{-1} for the myoglobin fluoride, and the results suggested that E was about 4 per cent of D. These measurements, however, were made at the relatively low microwave frequencies of 9000 Mc/s and hence the error in the values of D are likely to be rather high. Measurements by Ingram and Slade[14, 28] have been made at wavelengths of 4 mm, however, where significantly greater shifts of the $g_\perp^{\text{eff}}$ to 5·915 are obtained, and a detailed analysis of these results gives values of $2D = 8·8 \pm 0·8$ cm^{-1} for the acid-met derivative. Further measurements are now being made on other derivatives, and over different temperature ranges, so that a detailed picture of the effect of the internal molecular electrostatic field can be deduced.

It will be clear from all these measurements that considerable work still needs to be carried out in the electron resonance investigations of the different derivatives of myoglobin and haemoglobin before a complete understanding of the energy level system of the iron atom and its surrounding ligands is available, but the measurements made so far have indicated the great power of electron resonance to probe both the detailed structure and the molecular orbital configuration of the central part of these important biochemical molecules.

§6.7

ELECTRON RESONANCE STUDIES
ON NON-HAEM IRON IN PROTEINS

Although a large amount of interest in the binding of iron in bio-chemical compounds and in proteins in general has been concentrated on the iron atom in the haem plane, there are several other important biological molecules, in which iron is also present, but not in a haem-plane configuration. These molecules are also being investigated by electron resonance techniques, and although nothing like as much precise information on them has been obtained so far, as for the haem-containing proteins, it is nevertheless to be expected that much more information will become available as these studies are continued.

One general type of electron resonance observed from iron in proteins with a g-value of 1·94 has already been discussed in some detail in §5.7 in connection with the studies on enzymes containing iron atoms. It became clear in that section that the exact nature of the iron giving rise to this particular signal had not yet been elucidated and various different theories for its origin are still being discussed.

It should also be noticed that, under certain conditions, an electron resonance signal can be obtained from iron present in biological compounds with an effective g of 4·3. The existence of an absorption from iron atoms with this particular g-value was first noticed by Sands[29] in 1955, when making electron resonance investigations on glass. These can be explained in terms of iron atoms which are basically of the high spin variety with $S = \frac{5}{2}$, but are placed in a site with a large component of rhombic symmetry. The energy levels corresponding to the three pairs of spin states are then considerably admixed and, at the normal frequencies of measurement at X-band, the transitions occur at a field position corresponding to a g of 4·3. In the same way, just as the $g_\perp$ effective of the octahedral site is not actually 6·0, but only appears to be so under certain conditions, so this $g_\perp$ effective of 4·3 is really not a true g-value at all; in fact, more care must be taken when deducing conclusions from it than for the case of the $g_\perp^{\text{eff}}$ of 6·0. In other words, the energy levels in the fields with high rhombic components are changing rapidly with wavelength, and in each case a measurement of apparent g-value at different microwave frequencies is of considerable value. A detailed summary of the theory of this $g_\perp^{\text{eff}} = 4·3$ resonance has been given by Blumberg[30] and readers interested in the precise details of this resonance are referred to his article. He also ends up with the final conclusion that the amount of information obtainable on this type of iron rises rapidly the higher the microwave frequency used for the observation of resonance.

Shortly after this g-value of 4·3 was observed in the inorganic glasses, it was also reported in a variety of biochemical compounds,[31, 32] and biological specimens,[33] and a good example of

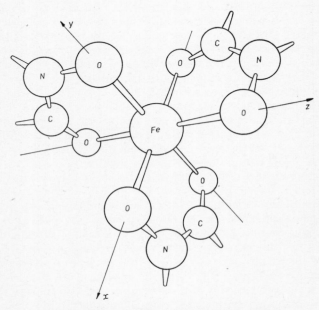

FIG. 6.26. Surroundings of iron atom in ferrichrome

The Fe^{3+} ion is surrounded by an octahedron of oxygen atoms, but these bind to form closed loops which severely distort the octahedral symmetry and produce a large rhombic component

this is the work of Wickman *et al.*[33] on the electron resonance of ferrichrome. Ferrichrome is a cyclic hexapeptide, obtained from the fungus *Ustillago sphaerogena*, and the detailed X-ray studies on this have shown that the iron atom is surrounded by an octahedron of oxygen atoms. The iron is thus sitting in a site of octahedral symmetry to a first approximation, but the binding to these six oxygen atoms produces a considerable rhombic distortion, as shown in Fig. 6.26. Rather wide electron resonance absorption lines are obtained from these iron atoms, but centred on a g-value approximately equal to 4·3. Wickman *et al.*[33] show that these

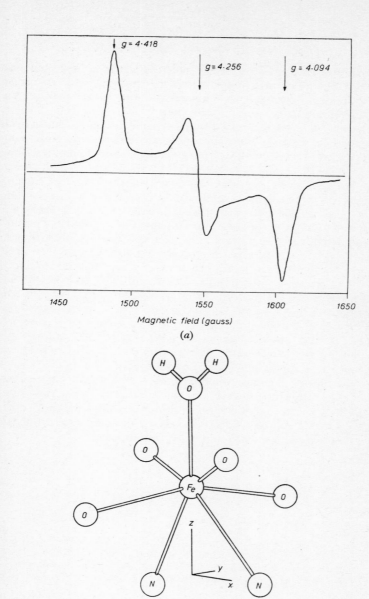

FIG. 6.27. E.S.R. spectrum for Fe^{3+} —EDTA complex

The three distinct turning points in the spectrum (*a*) correspond to the three different principal *g*-values, indicating considerable rhombic distortion. The actual surroundings of the Fe^{3+} ion are shown in (*b*), and the asymmetry in these can be clearly seen (From Aasa *et al.*[34])

measurements can be explained in terms of the rapidly changing energy levels associated with the highly rhombic field, and these measurements correlate very nicely with the X-ray work on the same compound.

Similar resonances with a g-value centred on 4·3 have also been obtained on other biologically important compounds containing iron, and as an example the spectra obtained from a polycrystalline Fe^{3+} — EDTA complex, as observed by Aasa et al.[34] is shown in Fig. 6.27. It can be seen that this curve has three distinct turning points and would thus normally be interpreted as corresponding to a g-value variation with three different principal axes; the assignment of these three g-values is shown in the figure. The surroundings of the iron atom in this particular case are illustrated in Fig. 6.27(b), and again it will be seen that although there is octahedral symmetry to a first approximation, there will be very considerable rhombic distortion as well. Blumberg[30] has shown that these resonance results can be interpreted in terms of splitting parameters $D = 0.769$ cm^{-1} and $E = 0.236$ cm^{-1}, which agree well with the theoretically predicted values for this particular configuration. Similar measurements on other biochemical compounds containing iron in these sites of low symmetry can also be analysed by the general method developed by Blumberg. These results serve to show how even complex configurations around the central paramagnetic atom with relatively low symmetry, can still be studied in detail by the electron resonance technique. Moreover, these measurements can still be made and analysed in detail even when the specimens are in a polycrystalline, or liquid, state.

References

1. Beringer, R., and Rawson, E. B., *Phys. Rev.*, 1952, **87**, 228, and 1952, **88**, 677.
2. Beringer, R., and Heald, M. A., *Phys. Rev.*, 1954, **95**, 1474, and 1954, **96**, 645.
3. Orgel, L. E., *Ligand Field Theory* (Methuen, London, 1960) and Griffith, J. S., *Theory of Transition Metal Ions* (C.U.P., London, 1961).
4. Malmstrom, B. G., and Vanngard, T., *J. Molecular Biol.*, 1960, **2**, 118.
5. Malmstrom, B. G., *Oxidases and Related Redox Systems* (Wiley, New York, 1966).
6. Vanngard, T., and Aasa, R., *Paramagnetic Resonance*, Vol. II, p. 509 (Academic Press, New York, 1963).
7. Malmstrom, B. G., and Vannguard, T., *J. Molecular Biol.*, 1960 **2**, 118.
8. Vanngard, T., *Magnetic Resonance in Biological Systems*, p. 213 (Pergamon Press, Oxford, 1967).
9. Peisach, J., Levine, W. G., and Blumberg, W. E., *Magnetic Resonance in Biological Systems*, p. 199 (Pergamon Press, Oxford, 1967).
10. Bleaney, B., and Ingram, D. J. E., *Nature*, 1949, **164**, 116.

11. Ingram, D. J. E., *Spectroscopy at Radio and Microwave Frequencies*, p. 169 (Butterworths, London, 1967).
12. Hogenkamp, H. P., Barker, H. A., and Mason, H. S., *Arch. Biochem. Biophys.*, 1963, **100**, 353.
13. Kramers, H. A., *Proc. Acad. Sci. Amst.*, 1930, **33**, 959.
14. Slade, E. F., and Ingram, D. J. E., *Nature*, 1968, **220**, 785.
15. Bennett, J. E., Gibson, J. F., and Ingram, D. J. E., *Proc. Roy. Soc. A.*, 1957, **240**, 67.
16. Kendrew, J. C., Parrish, R. G., Marrack, J. R., and Orleans, E. S., *Nature*, 1954, **174**, 946.
17. Perutz, M. F., *Proc. Roy. Soc. A.*, 1954, **225**, 264.
18. Bennett, J. E., Gibson, J. F., Ingram, D. J. E., Haughton, T. M., Kerkut, G. A., and Munday, K. A., *Physics in Medicine and Biology*, April, 1957, p. 4.
19. Kendrew, J. C., and Parrish, R. G., *Proc. Roy. Soc. A.*, 1956, **238**, 305.
20. Stryer, L., Kendrew, J. C., and Watson, H. C., *J. Molec. Biol.*, 1964, **8**, 96.
21. Bennett, J. E., Gibson, J. F., Ingram, D. J. E., Haughton, T. M., Kerkut, G. A., and Munday, K. A., *Proc. Roy. Soc. A.*, 1961, **262**, 395.
22. Stryer, L., Kendrew, J. C., and Watson, H. C., *J. Molec. Biol.*, 1964, **8**, 96.
23. Morita, Y., and Mason, H. S., *J. Biol. Chem.*, 1965, **240**, 2654.
24. Helcke, G. A., Ingram, D. J. E., and Slade, E. F., *Proc. Roy. Soc. B.*, 1968, **169**, 275.
25. Kotani, M., and Morimoto, H., *Magnetic Resonance in Biological Systems*, p. 135 (Pergamon Press, Oxford, 1967).
26. Ingram, D. J. E., and Slade, E. F., *Proc. Roy. Soc. B.*, 1969 (in press).
27. Eisenberger, P., and Pershan, P. S., *J. Chem. Phys.*, 1967, **47**, 3327.
28. Ingram, D. J. E., and Slade, E. F., *Proc. Roy. Soc. A.*, 1969 (in press).
29. Sands, R. H., *Phys. Rev.*, 1955, **99**, 1222.
30. Blumberg, W. E., *Magnetic Resonance in Biological Systems*, p. 119 (Pergamon Press, Oxford, 1967).
31. Windle, J. J., Wiersema, A. K., Clark, J. R., and Feeney, R. E., *Biochemistry*, 1963, **2**, 1341.
32. Aasa, R., Malmstrom, B. G., Saltman, P.. and Vanngard, T., *Biochim. Biophys. Acta.*, 1963, **75**, 203.
33. Wickman, H. H., Klein, M. P., and Shirley, D. A., *J. Chem. Phys.*, 1965, **42**, 2113.
34. Aasa, R., Carlsson, K. E., Reyes, L. S. A., and Vanngard, T., *Arkiv. Kemi*, 1966, **25**, 285.

Chapter 7

Recent Developments
and
Future Prospects

§7.1

THE APPLICATION
OF SATURATION AND DOUBLE RESONANCE STUDIES
TO BIOCHEMICAL COMPOUNDS

All the electron resonance studies on biochemical or biological compounds that have so far been summarized in detail in this book have followed along the lines of investigation and analysis that were developed in the general field of electron resonance over the past ten years or so. In this kind of analysis the actual electron resonance absorption is first recorded and then three basic parameters are deduced from it, i.e. (i) its integrated intensity, (ii) the associated g-values, and (iii) any hyperfine splittings present. These are then used qualitatively to characterize the entity being studied and also quantitatively to measure its concentration.

Thus the concentration of free radicals or other intermediates, in enzyme reactions or other similar processes, is determined by measuring the integrated intensity of the absorption line and plotting it against time, in order to establish the overall kinetics of the system. The actual nature of the free radical intermediate can then be deduced from the hyperfine structure that is observed on the spectrum. Any metal ions that are taking place in the reaction, or are present in the biochemical molecule of interest, can also be identified in this way or from the g-value variation, which can be used, not only to identify the ions themselves, but also to give very significant information on the nature of their chemical bonding and coordination within the molecule. These are, in fact, the standard methods employed in the analysis of any normal electron resonance spectrum, and the way in which the analysis is carried out is now very well established both in theory and in practice.

There are other ways, however, in which considerable additional information can often be obtained from electron resonance studies and which are now beginning to be applied to biochemical and biological specimens. One group of methods can be brought together under the general heading of 'saturation studies'. They are essentially concerned with a measurement of the relaxation processes, and times, of the paramagnetic group within the molecule. The methods normally investigate the variation of the resonance line-widths under different input-power conditions. Such studies on relaxation times and mechanisms can often give very helpful additional information on the bonding of the metal ions or free radical components and on their interactions with other atoms within the molecule. The general theory of saturation broadening of electron resonance lines has already been discussed in some detail in §3.7; and the way in which relaxation times can be measured experimentally by such saturation studies was outlined in §3.8. The application of these ideas and techniques to typical biological compounds is therefore taken up in the next section to illustrate how such studies are now helping in the elucidation of biochemical structures.

One other more specialized technique which has also been developing in the general field of electron resonance over the last few years is that of double resonance, in particular the ENDOR technique. The basic principles and methods of this technique have already been discussed in §3.9, where it was seen that the crucial contribution that the technique can make is to resolve very small hyperfine splittings in broadened electron resonance lines. In this way very precise information can be obtained on both the presence and the magnitude of these interactions. Although the initial studies were all made on single crystals, the technique has now been applied to solutions and hence is becoming more readily available for the study of biological and biochemical specimens. Some examples of the application of this kind of investigation are given in the section following that on saturation studies.

Apart from the application of these newer methods, which have been developed in the general field of electron resonance, there have also been some recent developments specific to the field of biochemical investigation itself. One of the most striking of these is known as the 'spin-labelling technique', in which certain well-characterized paramagnetic radicals are grafted on the biochemical molecules being investigated, and the orientation, in which this free radical marker is then held, can be determined from the electron resonance spectra. By grafting the radical on different parts of the biochemical molecule, the general orientation of different groupings within the molecule can be determined, as well as any magnetic interactions within the molecule being studied. Examples

of the way in which this technique is being applied to various
biochemical specimens, including such molecules as haemoglobin,
is outlined in the fourth section of this chapter.

§7.2

RELAXATION STUDIES ON BIOCHEMICAL MOLECULES

It will be remembered from the general introductory sections, and
also from §3.7 in particular, that there must be some relaxation
processes present within any molecule, in order to allow the energy
absorbed in the electron resonance transitions to return to the
ground state. This also allows continuous absorption of the incoming microwaves to take place, and continuous observation of
the electron resonance signal is thus possible. If the incoming
microwave power becomes large, however, these relaxation processes may not be able to return the energy to the ground state
sufficiently rapidly, and under these conditions 'saturation' of the
resonance will take place. The detailed expression governing this
saturation was derived in equation (3.11). It can be seen from that
equation that there are, in essence, two parameters to be considered; one is the spin lattice relaxation time, T_1, which is usually
the constant to be determined, and the other is the value of the
applied microwave magnetic field, H_1, which is normally the parameter experimentally varied in order to determine T_1. Thus the
magnitude of the incoming microwave power can be increased,
and hence $H_1{}^2$ increased, until a measurable amount of saturation
takes place. This amount of saturation can then be measured
quantitatively and used to determine the actual magnitude of T_1
from equation (3.11).

It is clear that the experimental techniques in such a measurement are basically the same as those in a standard electron resonance spectrometer, except that a high power source is required as
well as methods of rapidly monitoring both the intensity and width
of the absorption line. The way in which this is carried out in practice has already been discussed in some detail in §3.8, and an
example of a spectrometer constructed specifically to measure
relaxation times in this way has been given in Fig. 3.24. The
measured decay of a typical electron resonance signal, following
the incoming microwave pulse as produced by such a spectrometer, was also shown in Fig. 3.25. This section will therefore be
confined to a consideration of the actual results obtained when
these techniques are applied to biochemical compounds, together
with a brief discussion of the implications of these measurements.
One good example of the way in which different components of an

electron resonance spectrum can be differentiated by relaxation measurements is afforded by the work on aldehyde oxidase, and in particular the study of the molybdenum metal ions in this enzyme. Aldehyde oxidase can be obtained from rabbit liver and has a structure and properties[1] which are very similar to those of xanthine oxidase, which was discussed in considerable detail in Chapter 5. It contains both molybdenum and iron, as in the case of xanthine, having an approximate molecular weight of 280 000 with two atoms of molybdenum and eight atoms of iron per molecule.[2]

It will be remembered that the Mo(V) in the xanthine oxidase had a signal with a characteristic g-value of 1·97, and this was also found in the aldehyde oxidase. It is present, moreover, in the normal isolated condition of the enzyme, corresponding to its oxidized, or 'resting', state. The six-line hyperfine pattern due to the Mo^{95} and Mo^{97} isotopes, can also be identified on this signal, confirming its origin from the molybdenum atoms. As mentioned earlier the amount of Mo(V) in this resting condition corresponds to about 25 per cent of the total amount of Mo(V) observed by electron resonance, even when the enzyme is active. Hence it might appear that it would be very difficult to differentiate between the molybdenum taking an active part in the enzyme reaction from that present all the time in the resting enzyme, since these two signals will be directly superimposed one on top of the other. However, it has been found[3] that the Mo(V) signal produced on reduction during the enzyme activity has a shorter relaxation time than the signal from the resting enzyme. It is therefore possible to measure the ratio of these two signals present in an accurate way, by determining the change in the composite relaxation time, as the enzyme activity grows. This change of measured relaxation of the Mo(V) signal towards lower values, as the enzyme activity progresses has, in fact, been experimentally observed,[4] and it has thus been demonstrated that this differentiation by relaxation time measurements is a practical possibility in such cases as these.

This differentiation between the two types of Mo(V) ion present in the aldehyde oxidase enzyme, has also been employed to sort out the kinetics of the electron transfer sequences during the initial stages of enzyme activity.[4,30] Thus, when the enzyme is titrated anaerobically with limited amounts of substrate, reducing equivalents are initially stored up in the flavin and iron components, whereas little change is seen in the Mo(V) signal of the resting enzyme. The signal from the free-radical species goes through a maximum at about 3 electron equivalents per flavin, and, at the same time, approximately one half of the iron signal at $g = 1·94$ appears. However, it is only after the addition of 6 electron equivalents per flavin that the Mo(V) signal suddenly changes its relaxation time, and the component characteristic of the molybdenum associated with the enzyme activity appears.

This fact is illustrated diagrammatically in Fig. 7.1, where the different curves from the free radical signal, the iron, and the molybdenum, are plotted against the number of electron equivalents

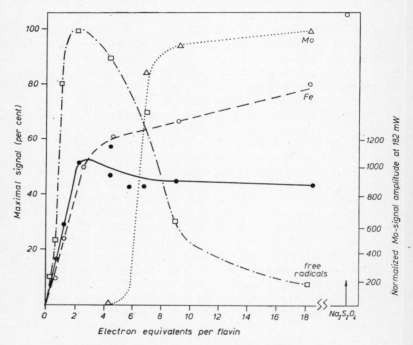

FIG. 7.1. Kinetics of enzymatic activity of aldehyde oxidase

The variation in intensity of the different E.S.R. signals for the free radicals, the iron and the molybdenum, are all plotted against the electron equivalents per flavin. The intensities are plotted as a percentage of the maximum attained, and the input power was 0·5 milliwatt. The saturation effect on the molybdenum signal can be seen from the full curve, obtained for a microwave power of 180 milliwatt (From Rajagopalan *et al.*[30])

added per flavin. The vertical axis corresponds to the signal amplitude, plotted as a percentage of the maximum amplitude observed for each type of signal, and those for the molybdenum were corrected for the amplitude of the signal found in the resting enzyme. These particular curves were obtained at a microwave power level of 0·5 milliwatt. In contrast the heavy full line indicates the curve that is obtained from the molybdenum signal if a microwave power of 182 milliwatt is used. The vertical scale for this signal is shown on the right of the figure, but it is clear from the completely different shape of this variation that the two types of Mo(V) present have very different relaxation times.

Beinert and Orme-Johnson,[4] have interpreted these relaxation measurements on Mo(V) as an indication that the molybdenum initially present in the enzyme increasingly interacts with the paramagnetic species generated in its vicinity, i.e. the flavin semiquinone and the iron component corresponding to the g of 1·94. Maximum interaction then occurs when the maximum amount of semiquinone is present, but since the $g = 1·94$ signal keeps on increasing, as the semiquinone signal disappears, compensation for this disappearance is provided. Furthermore, detailed studies of the kinetics of these changing relaxation times have allowed a more precise elucidation of the different interactions taking place between the various paramagnetic species in this enzyme. Those interested in these finer details are referred to a review article by Beinert and Orme-Johnson[4] on this topic.

The last paragraphs have shown how saturation studies can be used to distinguish between different paramagnetic species in a reacting enzyme, and this method of differentiating one metal ion from another has obviously a large number of potential applications in the field of biochemistry. It is also possible, however, to use saturation studies and measurement of relaxation times to probe the actual interactions that are taking place within the molecules themselves. An example of this work is that undertaken by Beinert and his colleagues[5] on the effect of adding nickel ions to various flavin derivatives. Earlier work[6] on free flavin semiquinones showed that they formed more stable metal chelates at their fourth and fifth coordination positions than do either the oxidized or fully reduced flavins. It was also found that these complexes of paramagnetic metals with the flavin semiquinones gave no E.S.R. signals, and thus either they were diamagnetic or they had signals broadened beyond detection.

Further study[7] of these reactions then showed, however, that if short-term oxidations of such derivatives as 5 benzyl -1, 3-dimethyl leucolumiflavin were carried out, so as to avoid decomposition of the flavin (and this could be checked by spectrophotometry), then added nickel ions would produce an enhancement of the relaxation similar to the case of the unblocked flavins. An example of these kind of observations is shown in Fig. 7.2. This figure shows the change in saturation produced in the lumiflavin semiquinone, on the addition of the nickel ions. Thus the lower curve is the one for the electrolytically oxidized flavin in the absence of nickel ions, while the upper curve is the same as the lower one, but with nickel chloride added at a concentration of 5×10^{-2} molar. In this figure, the saturation (or more precisely the logarithm of the signal amplitude divided by the square root of the power), is plotted as the vertical axis, while the logarithm of the power itself is plotted horizontally.

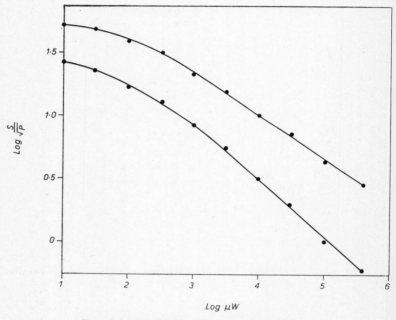

FIG. 7.2. Effect of nickel ions on saturation
of dimethylbenzyl-lumiflavin semiquinone

The lower curve is for the lumiflavin semiquinone formed by electrolytic oxidation in the absence of the nickel ions, while the upper curve is for the same conditions as the lower curve, but with 5×10^{-2}M of $NiCl_2$ added after oxidation (From Beinert *et al.*[7])

Further studies[7] on these blocked flavins in other media led to the general conclusion that interactions between flavins and metals might occur in other ways than chelation. One such possible interaction, which has been suggested, is the formation of a 'bridge complex', in which an anion acts as an intermediary between the cationic flavin semiquinone and the paramagnetic metal ion. This complexing anion may then furnish a relaxation mechanism between the flavin and paramagnetic metal. Although further work must be done to confirm the exact nature of such interactions as these, it is clear from such experiments that saturation studies, and measurements of relaxation times, promise to give very useful information on the detailed interactions between metal ions and the surrounding groups within the biochemical molecule itself.

§7.3

THE APPLICATION OF DOUBLE RESONANCE TECHNIQUES

The way in which the principles and techniques of ENDOR as described in §§3.8 and 3.9, can be applied to biochemical systems, is probably best introduced by discussing the measurements on a compound such as copper phthalocyanine. This has a structure very similar to some of the important biochemical compounds, and is one on which specific ENDOR measurements have been made. This particular molecule was considered as an example of superhyperfine splitting in the introductory sections, and its structural form is shown in Fig. 1.14(a). It is seen there that the central copper atom is surrounded by a square of four nitrogen atoms and the electron resonance spectra show that the wave function of the magnetic electron on the copper atom spreads out to include these nitrogen atoms, since a superhyperfine pattern of nine lines is observed superimposed on each of the copper hyperfine absorptions. This superhyperfine pattern has been illustrated in Fig. 1.14(b). A detailed analysis of the splitting of this nitrogen superhyperfine system can give very precise information on the molecular orbital associated with the copper atom and its surroundings.

Thus the energy-level system for the complex can be built up in the way already discussed in §6.3 and as illustrated in Fig. 6.6. It was seen there that the unpaired electron resides in the $B_{1g}{}^*$ antibonding orbital, with the $B_{2g}{}^*$ and $E_g{}^*$ antibonding orbitals immediately above it. It can also be seen that the $B_{1g}{}^*$ orbital has been formed by an admixture of $d_{x^2-y^2}$ orbital from the central paramagnetic ion, together with admixture from the sigma bonds of the ligand atoms. The expression for this orbital can, in fact, be written as

$$\psi_{B_{1g}{}^*} = N \cdot d_{x^2-y^2} - \frac{\lambda_s}{2}(-s_1 + s_2 - s_3 + s_4) - \frac{\lambda_p}{2}(-p_1 + p_2 - p_3 + p_4)$$

$$(7.1)$$

where N measures the fractional admixture from the central copper atom, while λ_s and λ_p measure the spread of the wave function on to the surrounding nitrogen atoms. The s_1, s_2, s_3, s_4 and p_1, p_2, p_3, p_4 represent the $2s$ and $2p$ orbitals on the four surrounding nitrogen atoms.

The theoretical expression for the g-values of the complex can be derived in the way outlined in §6.3 and these expressions will also contain admixture coefficients for the higher $B_{2g}{}^*$ and $E_g{}^*$ orbitals. The magnitude of the superhyperfine splitting observed for the nitrogen atoms can also be employed, however, to give a direct determination of the values of λ_s and λ_p in the expression for the wave function of the $B_{1g}{}^*$ orbital. The magnitude of the superhyperfine splitting parallel to the axis of symmetry, through the copper atom can in fact, be expressed as[8]

$$A = \frac{16\pi}{3} . \gamma . \beta . \beta_N \left| S(0) \right|^2_{2s} . \frac{N^2 . \lambda_s{}^2}{4} + \frac{8}{5} \gamma . \beta . \beta_N . \left(\frac{1}{r^3} \right)_{2p} . \frac{N^2 . \lambda_p{}^2}{4}$$

$$(7.2)$$

and a somewhat similar expression can also be derived for the splitting in the perpendicular direction. It is clear that a very direct measure of the coefficients λ_s and λ_p is thus available from such measurements of superhyperfine splittings, since these can be determined much more accurately by ENDOR than from simple E.S.R. splittings themselves. This is one way in which ENDOR can provide additional very precise and direct quantitative data on chemical bonding.

Copper phthalocyanine is one of the metallo-organic compounds on which ENDOR measurements have been made,[8] and the superhyperfine splitting has been checked directly in this way. The actual ENDOR absorption obtained for copper phthalocyanine is shown in Fig. 7.3, and as explained in §3.9 the horizontal axis now represents the nuclear resonance frequency and the vertical axis measures the desaturation of the electron resonance line. The magnitude of the superhyperfine splittings is thus obtained directly from the actual resonance frequency at which the desaturation takes place, and not from any splitting between the peaks of this ENDOR curve. The magnitude of the superhyperfine splitting measured in this way[8] is equal to 42 Mc/s and substitution of this figure into equation (7.2) gives values for λ_s and λ_p which agree well with the expected $s\,p^2$ hybridization of the ligand bonds.

It will be appreciated that the structure of the phthalocyanine molecule is very similar to that of the haem plane of haemoglobin and other related biochemical compounds. It would therefore be expected that double resonance ENDOR techniques might also give precise information on the spread of the wave function out from the central iron atom to the nitrogens in this case as well. ENDOR measurements in fact have been attempted on myoglobin and haemoglobin crystals, but so far no direct measurement of the superhyperfine splittings from the nitrogens has been observed. This is probably due to the broadening which is produced by the random orientation of the molecular axes and further experiments at different temperatures may help to elucidate this. However, some double resonance results have been possible and the interaction with the protons in the surrounding haem plane has been observed by Eisenberger and Pershan.[9]

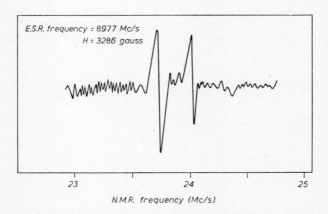

FIG. 7.3. ENDOR spectra of copper phthalocyanine
The horizontal axis represents the changing nuclear resonance frequency, while the vertical axis shows the desaturation of the electron resonance absorption

Although all the initial ENDOR experiments were carried out with single crystals, it has been possible in recent years to apply the technique also to free radicals in solution.[10] In all ENDOR experiments it is important to optimize the experimental conditions, since saturation, or partial saturation, of the electron resonance absorption is one of the essential requirements. In the case of studies in solution very high microwave power may be needed for saturation, and it is normally found that the best results are obtained at temperatures just above the freezing point of the solvent and with concentrations of about 10^{-3} molar.[11] It is then sometimes possible to obtain significant saturation at microwave powers of about 1 mW incident on the cavity, and it is desirable to use solvents which have a low dielectric loss so that the maximum

T

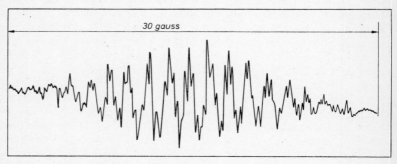

(a)

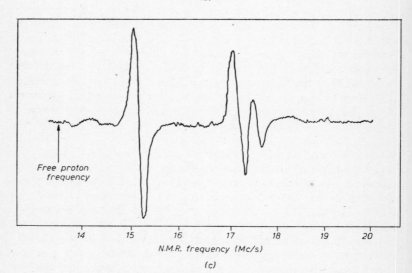

Fig. 7.4. E.S.R. and ENDOR spectra of triphenyl methyl
(a) Structural formula for triphenyl methyl
(b) E.S.R. spectrum under high resolution
(c) ENDOR spectrum showing that only three basic interactions are present (From Hyde[11])

filling factor can be used. In most ENDOR experiments on saturation, however, it is wise to have high microwave input power available,[12, 13] since the relaxation times can become very short well away from the melting point.

One of the great advantages of employing ENDOR measurements can be very strikingly demonstrated when the spectra of free radicals are considered. It will be remembered that there are often two or more sets of carbon atoms in the aromatic ring structure, having different spin densities associated with them. These then give rise to various sets of unequal hyperfine splittings, which combine to produce a very complex structure. As a particular case in point, the spectra of triphenylmethyl is shown in Fig. 7.4, its structural formula is shown in Fig. 7.4(*a*) and the observed E.S.R. spectrum in Fig. 7.4(*b*). Although this spectrum contains a large number of lines, which are difficult to analyse immediately, they are all produced by three basic splittings. In the ENDOR technique these basic splittings are measured directly, instead of the complex pattern which each produces, and the ENDOR spectrum of triphenylmethyl,[11] as plotted against the nuclear resonance frequency, is shown in Fig. 7.4(*c*). It can be seen from this that the actual magnitudes of the splittings can be read off immediately from the observed ENDOR lines, and this simple spectrum contains all the information that is also present in the E.S.R. spectrum itself; and the analysis of the wave function is thus made very much more simple and direct. This applies just as much to the more asymmetrical molecules, as the example in Fig. 7.5 shows. In this case, a $CH_2 - SCH_3$ group has been added to the triphenylmethyl, as shown in Fig. 7.5(*a*); the ordinary electron resonance spectrum from this derivative is again extremely complex with lines of slightly varying intensity, as indicated in Fig. 7.5(*b*). The ENDOR spectrum of this derivative is shown in Fig. 7.5(*c*), and although this now contains more peaks than the ENDOR spectrum of the original triphenylmethyl, it is still very much more simple to measure the splittings directly from this ENDOR spectrum, than to attempt a complex analysis of the electron resonance spectrum itself. It was, moreover, found that, when the triphenylmethyl derivative was raised in temperature through $-20°C$, a significant change in the ENDOR spectrum was observed, showing that a conformational change takes place. Thus, above $-20°C$, several pairs of lines collapse to take on average values, showing that exchange is taking place between two conformations of this molecule at the higher temperature. The kind of changes that this effect produces in the electron resonance spectrum itself are extremely hard to analyse unambiguously, and this again illustrates the power

of ENDOR measurements to follow changes in hyperfine interactions very directly.

As an example of the application of these ENDOR studies on
free radicals which are of biological interest, the work of Hyde[11]
on the lumiflavin radical can be quoted. He studied the lumiflavin

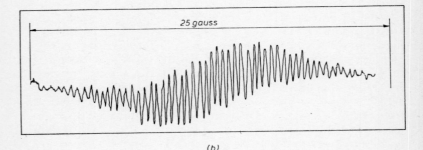

(a)

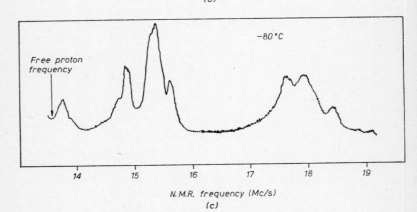

FIG. 7.5. E.S.R. and ENDOR spectra
of asymmetrical triphenyl methyl derivative

(*a*) Structural formula for triphenyl methyl derivative
(*b*) E.S.R. spectrum (From Hyde [11])
(*c*) ENDOR spectrum showing the ten different interactions now
present (From Hyde[11])

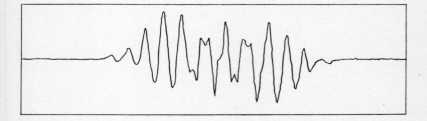

(a)

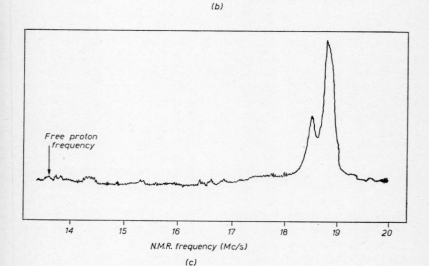

(b)

(c)

FIG. 7.6. E.S.R. and ENDOR spectra
of zinc chelated lumiflavin radical

(*a*) Structural formula, showing $CH_2\,COO\,C_2H_5$ group and Zn atom
added to the basic lumiflavin structure
(*b*) E.S.R. spectrum (From Muller *et al.*[14]) with 99% Zn^{64}
(*c*) ENDOR spectrum, showing the two main interactions (From
Hyde[11])

radical which had been chelated with zinc and which has the structural formula as shown in Fig. 7.6(*a*). The straightforward electron resonance spectrum of this radical had already been studied by Muller, Ehrenberg and Eriksson,[14] and is shown in Fig. 7.6(*b*). The fourteen lines observed in this spectrum can be interpreted as arising from equal couplings of about 3·5 gauss to the methyl protons at positions 7 and 9, as shown in Fig. 7.6(*a*), and to the proton at position 5 and to the nitrogens at positions 9 and 10. The ENDOR spectrum which Hyde observed from this same radical in solution is shown in Fig. 7.6(*c*), and it can be seen immediately that only two resonances are observed, one at 18·843 and the other at 18·526 Mc/s. These frequencies correspond to magnetic field couplings of 3·72 and 3·50 gauss, thus agreeing with the observed hyperfine splitting in the E.S.R. spectrum of about 3·5 gauss. Again it is quite clear that the ENDOR affords a very much more direct method of measuring the hyperfine couplings. This is also illustrated by recent ENDOR studies on dehydrogenase.[29]

As a link between the ENDOR measurements on free radicals in solution, and those on single crystals of biochemical compounds, it may be mentioned that experiments are now in hand[11] to carry out ENDOR measurements on spin-labelled biochemical molecules, as are described in the next section. It will be seen in that section that significant information can be deduced from the electron resonance spectrum which is obtained from specific free radicals grafted on to various parts of biochemical molecules. ENDOR measurements on these spin labels should therefore increase the precision of this information, in just the same way as it has for the simple free radicals themselves. These studies of both ENDOR on biological molecules and the spin labelling technique itself are very much in their early stages, but it is likely that considerable advance will take place in both fields of investigation in the near future.

Although not specifically included within the heading of ENDOR measurements, other types of electron-nuclear double resonance on biochemical and biological molecules are now being made and should be mentioned. Thus it is possible to carry out more or less the inverse of the ENDOR technique and to observe the nuclear resonance absorption from biochemical molecules and the effect that the presence of paramagnetic systems have on this nuclear resonance absorption. Recent work in this direction,[15] has shown that the spin lattice relaxation rate of the water protons and others bound to biological material can be very significantly affected by the presence of unpaired electron spins. When such cross coupling is present, it is possible to enhance the nuclear resonance absorption by several orders of magnitude, and this type of double resonance experiment, which is broadly classified as an Overhausser[16] effect, has been finding increasing importance in nuclear resonance studies. Although such experiments really come within the field of a text on nuclear resonance, rather than one on electron resonance, the existence of this other method, in which electron-nuclear cross interactions can be probed, should be known to all those interested in study of such relaxation mechanisms in biochemical compounds.

§7.4

SPIN LABELLING TECHNIQUES

It has already been mentioned in the introductory section of this chapter that the technique of 'spin-labelling' is one that has been developed specifically for the study of biochemical and biological molecules. It is thus distinct from all the other electron resonance techniques, which were developed initially, and applied for some years, in the general field of E.S.R. before being taken over into the biochemical field of study. The great advantage of spin-labelling is that it enables a well-characterized parmaganetic entity to be attached to a large biochemical or biological molecule, and the new environment of the spin-label itself can then be studied in detail. Both the orientation and the nature of chemical bonding around the spin-labelling free radical can be investigated, and the presence and onset of molecular motion can also be probed very precisely. It thus promises to be a technique with a large number of applications in the biochemical and biological field. At the moment, its possible applications are only just beginning to be appreciated.

The technique was really developed by McConnell,[17, 18] who initiated this idea of attaching a stable synthetic paramagnetic organic free radical to molecules of biological interest and has proceeded to study quite a large number of single crystals of proteins, which have been spin labelled in this way. Various different free radicals can be employed as the spin labels, but they should possess one or two essential features, such as high stability, and an electron resonance spectrum which varies with orientation of the radical, so that the orientation which the radical takes up in the biochemical molecule can be identified from the observed electron resonance absorption. McConnell and his co-workers[17, 18] found that free radicals which contain the nitroxide group with a nitrogen atom bound to tertiary carbon atoms formed one of the most successful types of spin-label.

The structure of such a free radical is shown in Fig. 7.7(*a*) and the R and R* represent groups that are used to bind the radical to the biochemical molecule. This binding may be covalent, hydrophobic, or of other types, and depends on the nature of the R group. One of the advantages of this particular series of radicals is

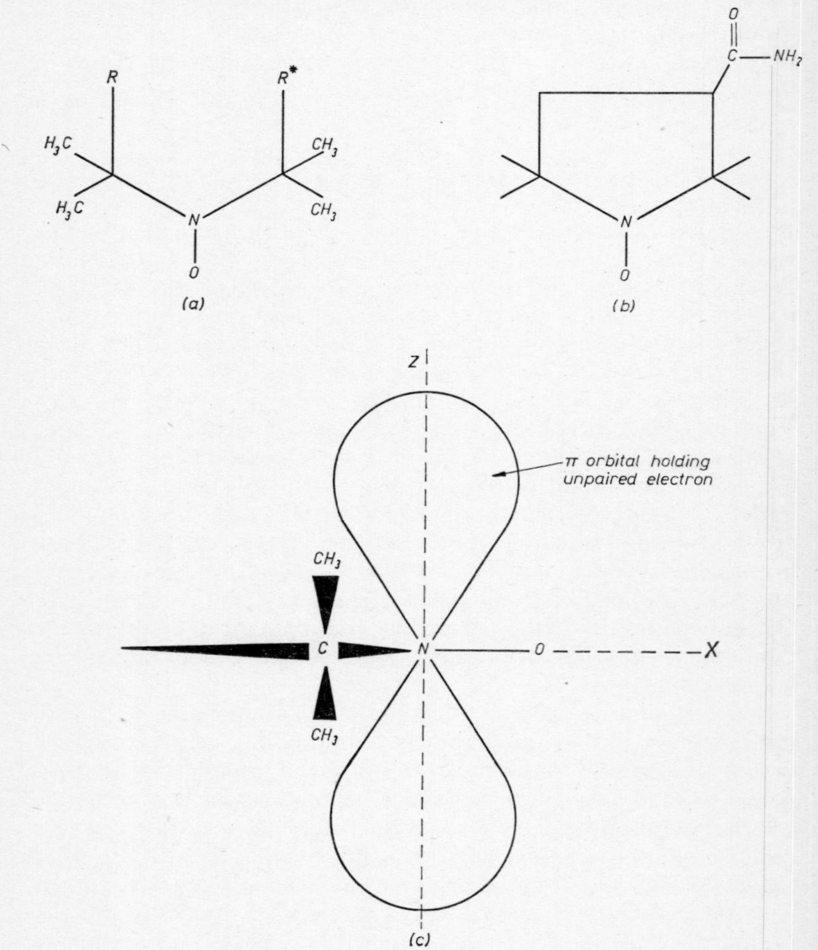

Fig. 7.7. Structure of typical 'spin-label'

(*a*) Basic structural formula, R and R* can be a variety of different groups and are used to bind the spin-label to the protein

(*b*) Examples of specific spin-label in which the two R groups join in a ring system

(*e*) Unpaired electron density over spin-label bonds

that a very large number of different groups, R or R*, can be used, and hence the spin label can be attached to a large variety of amino-acid side chains or other similar sites in the proteins. One of the main problems in this kind of work is to select the best type of group for R or R*, for the particular biochemical molecule that is to be studied, so that the spin label itself is attached at the most appropriate and informative points. In practice most of the spin labels so far used, have had R and R* groups that have been connected together to form 5- and 6-membered saturated rings, such as those illustrated in Fig. 7.7(*b*).

The wave-function that will be occupied by the unpaired electron associated with such a spin-label is represented schematically in Fig. 7.7(*c*). It can be seen that the major part of this will still be located in the *p*-orbit of the nitrogen atom of the nitroxide group, and this asymmetry in space is reflected by an anisotropy of the *g*-values which will have typical magnitudes[19] of

$$g_x = 2 \cdot 009$$
$$g_y = 2 \cdot 006$$
$$g_z = 2 \cdot 0027$$

The hyperfine splitting between the three lines from the nitrogen interaction are similarly anisotropic, and it is these well-characterized features of the spin-label which allows its orientation to be deduced from the observed E.S.R. spectra.

The actual binding between the biochemical molecule and the spin label itself can be sufficiently strong and rigid to prevent any motion of the nitroxide group relative to the rest of the molecule; alternatively, it can be quite weak and allow fairly free motion of the nitroxide group around bond axes, even when the rest of the molecule is immobile. These two different types of bonding, with a whole range of intermediates, give rise to significantly different electron resonance spectra, since those which contain nitroxide groups in rapid motion will give a spectrum similar to that observed from the simple spin-label by itself in solution, with lines narrowed by motional tumbling. On the other hand, biochemical molecules with immobilized nitroxide groups will produce electron resonance spectra with broad lines, similar to those observed from poly-crystalline or powdered samples of the spin-label. The actual width of the observed electron resonance lines thus gives some indication immediately of the nature of the bonding to the biochemical molecule, and the motion taking place within it.

It should be noted in this connection, that quite often the 'strongly immobilized spectra', a term taken to describe spectra similar to that observed from the solid and thus corresponding to

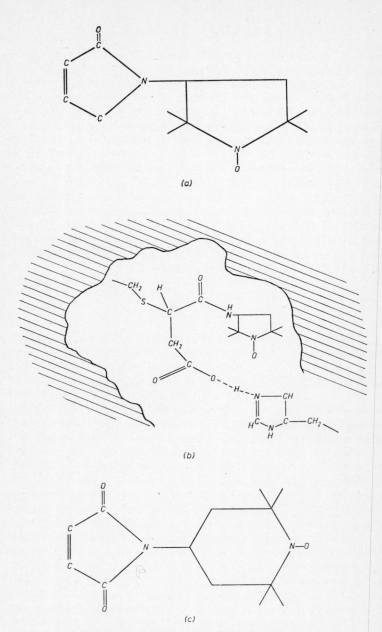

FIG. 7.8. Spin-labels on haemoglobin molecules
(a) Structural formula of spin-label initially used by McConnell
    et al.[20]
(b) Postulated opening of ring system by spin-label attachment
(c) Structural formula of improved maleimide spin-label

relatively little motion of the nitroxide group, are often observed from solutions when large protein molecules are being studied. This indicates that the protein molecule itself has a rigid internal structure and that its own tumbling is slow compared with the anisotropic energy terms in the spectrum. On the other hand, it is possible for narrow lines to be observed from the spin label, even when the biochemical molecule itself is not tumbling fast. This implies that the bonding of the spin-label to the rest of the molecule is such as to allow rotational freedom of the nitroxide group around one of its bonds to the rest of the protein. It follows from these general comments that care has to be taken in interpreting the spectra associated with the spin-label, and investigations under different temperature conditions can often be very helpful. It also follows that, if such care is taken and as many complementary experiments as possible are carried out, very considerable information can be obtained on the motion of the biochemical molecule itself and also on the binding within it at various specific points.

The best way to illustrate the kind of information which this spin-labelling technique can give is probably to take specific examples, and the case of the haemoglobin molecule may again serve in this way. McConnell and his colleagues[20] have been undertaking a series of investigations, employing different spin labels attached to horse oxy- and deoxy-haemoglobins. In their initial studies, the spin label shown in Fig. 7.8(a) was attached to the haemoglobin derivatives in solution, and it was then found that both strongly immobilized and weakly immobilized electron resonance signals were obtained. The strongly immobilized signals were about ten times stronger than the weakly immobilized signals, and it was shown that they arose from the attachment of the spin label to the reactive SH groups on the β chains of the haemoglobin. The immobilization indicates a definite stearic attachment and is probably associated with an actual opening of a ring system as indicated in the Fig. 7.8(b). Further measurements were then made on this strongly immobilized spin label in single crystals of the horse haemoglobin and also with an improved spin-labelling free radical of the structure shown in Fig. 7.8(c). The actual results of biological significance, which can be deduced from these studies, can be briefly summarized as follows.

In the first place, the orientation of the strongly immobilized spin label will give a direct measurement of the orientation of that part of the biological molecule to which it is attached, and this was shown to be the case for the haemoglobin crystals, where the two-fold molecular rotation symmetry axis was clearly evident in the electron resonance spectra. Slight changes in possible orientations associated with different amino-acid sites, can also be picked up very sensitively by the electron resonance spectra. Thus, there appeared to be evidence from the E.S.R. signals of the spin-labelled horse oxy-haemoglobin crystals that there are two slightly different possible orientations of the spin label when it is attached to the SH group at the β-93 site. The two principal axes for these orientations have a difference of 15°, and it has been tentatively suggested[20] that these may be due to possible isomeric orientations and possibly to two different kinds of protein known to be present in the horse haemoglobin.[21] Although a large amount of work needs to be done to elucidate the detailed meaning of these slight differences in electron resonance spectra, it is quite clear that a very sensitive method is now available by which such small conformational changes can be studied.

There is no reason, of course, why more than one spin label should not be attached to a given biochemical or biological molecule. In fact, some interesting additional information can be obtained if spin-labels are attached to two different amino acid groups, which are relatively close together in the same molecule. It should then be possible to observe either the dipolar interaction, or spin exchange, between the two unpaired electrons on these two spin-labels, and thus obtain precise information on the actual proximity and orientation of the two labelled amino-acid groups with respect to one another. It is clear that there are a large number of possible additions and modifications to the simple method initially employed. There is no doubt that the literature of the next few years will contain many examples of the way in which such spin-labels have been employed to give precise information on the structure, chemical bonding, and internal motion, of biochemical and biological molecules.

§7.5

FUTURE PROSPECTS

It has become increasingly evident in the last few sections that the work reported there is very much in its initial and preliminary stages and that many of the applications of such techniques as double resonance, and spin-labelling, to biochemical and biological molecules still have to be discovered. In this connection, it is worth noting the rapidity with which electron resonance techniques of all kinds are now being applied to biochemical and biological studies; it is quite clear that this may well prove to be one of the most fruitful fields in which electron resonance has been applied. Since these are early days in this general field of investigation, it is probably irresponsible and unwise to attempt detailed predictions on either the final form that these studies will take or the way in which they are most likely to develop over the next few years. It may be proper and helpful, however, to conclude this book with a few general comments on the kind of problems which electron resonance might well be able to tackle in the future and, at the same time, point out some of the limitations, as well as the potentialities of the technique.

In the first place, it is clear that the standard methods of observing and analysing electron resonance spectra will play an increasing part in studies of radiation damage. This is one technique which can probe straight into the damage centre itself and can watch both the initiation of such damage centres and the way in which secondary centres appear and migrate. Although unambiguous and precise characterization of the particular damage centre may be somewhat hampered at the moment by poorly resolved hyperfine splittings, this is undoubtedly a field in which better and better characterization will be possible as more and more spectra are recorded and analysed. This general background information is now being steadily accumulated and more definite interpretations of the spectra, such as that now associated with the protection mechanism of the sulphur atoms, will be increasingly available. In this connection, it is clear that electron resonance measurements on single crystals of biochemical molecules will always be of immense value, if only to check on the interpretation of the results obtained from solutions, or other conditions closer to those *in vivo*.

It is also clear that another line of research which will develop rapidly over the next few years is that associated with the study of enzymes. The particular advantage of electron resonance studies in the investigation of these systems has been outlined in some detail in Chapter 5. Some of the examples given there indicate the great power of electron resonance in probing various features of these important biochemical reactions. The particular advantages are associated with the ability of electron resonance to characterize the interacting atom, or groups, very specifically and to follow various steps in the reactions by the kinetics associated with each intermediate product. In particular, the work on xanthine oxidase has shown the different ways in which precise information both on the free radical activity and on the valency changes in the metal ions can be obtained. The extension of similar electron resonance studies to a large number of other enzyme systems will undoubtedly follow in the near future. These measurements on enzymes serve to illustrate most of the advantages of electron resonance in the study of biochemical systems, since all the newer techniques of continuous flow and rapid freezing can be applied to them and the more sophisticated theoretical analysis, involving both g-value calculations and hyperfine splittings, can also be employed, to obtain direct information on the bonding and nature of the metal ions within them.

This probing of both the immediate structure around the active paramagnetic ion, and also the detailed nature of their own chemical bonding, will also continue to be applied to many other metallo-organic molecules as well as the enzymes themselves. The work summarized in Chapter 6 on various copper and iron containing proteins, and on haemoglobin and its derivatives in particular, has illustrated the kind of information obtainable when single crystals of these biochemical molecules are available. The most important point to be brought out from these particular studies, however, is probably the great benefit to be obtained when there is an inter-action between different methods of investigation. This general theme, of bringing many different physical techniques and chemical methods to bear on biological specimens, has been responsible for the birth of a whole range of subjects now included under the general term of 'molecular biology', and electron resonance studies of biochemical molecules has taken a significant place amongst these innovations. In all these cases, however, it has not been just one single technique, and the information it has obtained, that has provided the crucial new steps in an understanding of biological structure or mechanism. It has usually been one or two new tech-niques together, or older techniques applied in a new way, that have opened up the new fields of interest. The combination of the electron resonance measurements on haemoglobin, together with the X-ray work that was going on at the same time, is probably a very good example of how these different techniques can interplay and assist one another. Thus, the real contribution of electron resonance studies of the future may well be as a complement to other techniques, so that, together, they can give a fuller under-standing of biochemical and biological molecules, rather than providing a complete analysis on its own. In this connection, one other new technique should be mentioned in closing, since this has already had some interaction with the electron resonance studies, and may well do so in an increasing measure in the future.

This technique is known as Mössbauer spectroscopy, and al-though the wavelengths it studies are well away from those used in electron resonance, in fact at completely the other end of the electro-magnetic spectrum, a brief description of its principles and methods may help to show the way in which its results are assisting in the interpretation of some of the electron resonance measurements themselves. In essence the final information gained from a Mössbauer experiment is the same as that gained by electron resonance, since the effect of the molecular or solid-state environment on the energy levels of the central paramagnetic atom is measured in both cases. In electron resonance these energy levels are often determined by direct measurement of transitions between them, but in Mössbauer spectroscopy they are determined as very fine splittings on the very much larger energy transitions associated with γ-ray emission and absorption. The important information for the molecular and solid-state physicists, however, is the actual magnitude of these fine splittings, since it is from these that the chemical state, and detailed bonding of the atom in which the nucleus resides, can be determined.

In most cases, it would be quite impossible to resolve this very fine splitting in the normal γ-ray spectra, since the recoil of the emitting or absorbing nucleus would completely mask the very fine energy differences corresponding to the solid-state interactions. The great contribution that Mössbauer himself made was to point out that, in certain cases, the recoil of the nucleus could be entirely absorbed by the surrounding lattice, or molecule, and hence the energy spread was dispersed over a very large number of nuclei rather than over one. This collective action so minimizes the recoil energy of the individual nucleus, that the energy width of its γ-ray line becomes extremely small and allows resolution of the very fine solid-state interactions. Unfortunately, the conditions required for such sharing of recoil energy with the lattice as a whole are rather specific and, as a result, only relatively few nuclei with suitable nuclear energy level systems can be used for such studies. One of these, however, is Fe^{57}, and it is rather fortunate that this particular nucleus is also an important one in a large number of biochemical molecules. So far only biochemical molecules containing iron have been studied by the Mössbauer technique, but since these include haemoglobin and its various derivatives,[22, 23] together with cytochromes,[24] peroxidases,[25, 26] and a large number of non-haem iron proteins,[27, 28] the field is not as limited as it might appear at first sight.

As in all forms of spectroscopy, a range of wavelengths will be necessary in the source, if absorption spectra in a given specimen are to be studied, and the obvious source for such γ-radiation of very precise wavelengths will be the Fe^{57} nucleus itself. Since the nucleus of the emitting source must be at the higher energy level initially, the source actually consists of a Co^{57} nucleus which is decaying, with a half life of 267 days, to an excited Fe^{57} nucleus.

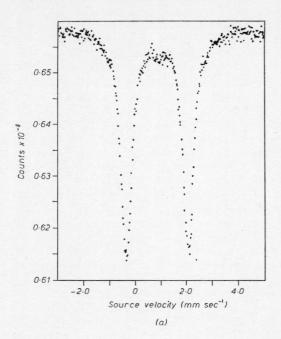

(a)

FIG. 7.9. (*Above and facing*) Mössbauer spectra from haemoglobin derivatives

The ordinates are counts $\times 10^{-6}$ in every instance

(a) Spectrum from reduced haemoglobin. The doublet splitting arises from the electrostatic interaction between the quadrupole moment of the Fe^{57} nucleus and the molecular electric field

(b) Spectrum from haemoglobin fluoride; the extra splittings are due to the high-spin ferric state

(c) Spectra from haemoglobin fluoride in an applied magnetic field of different strengths, from 7·5 kilogauss to 30 kilogauss. The extra splittings produced by this magnetic interaction can be clearly seen (From Lang and Marshall[22])

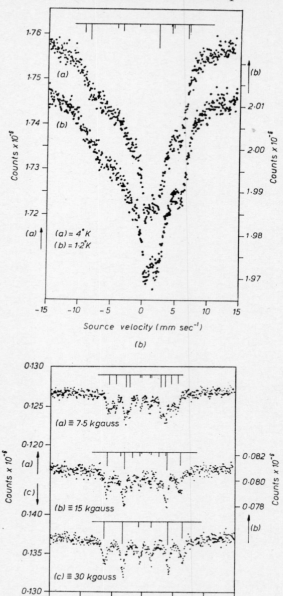

(b)

(c)

This excited Fe^{57} nucleus itself then decays to its ground state, emitting the appropriate γ-rays as it does so.

The actual range of wavelengths is then obtained by the very simple process of moving the Fe^{57} source at different speeds. Thus, the Doppler shift which is associated with any radiation, will ensure that a change of wavelength is produced, if the source itself is moving. Moreover this change will, to a first approximation, depend linearly on the rate of movement. Strange as it might appear at first sight, the Doppler speeds required are extremely small and the range of frequency which will cover all normal solid-state interactions can be produced by simply moving the source at speeds varying from ± 10 mm per second. This very small speed arises, of course, from the very high energy of the γ-ray and the very small fraction that the solid-state interactions are of this total energy.

It can be seen that the basic principles and methods of such Mössbauer spectroscopy are, therefore, relatively simple. The Co^{57}/Fe^{57} source, is mounted in such a way that it can have a velocity range within that indicated, and the absorption spectra of these γ-rays, as they pass through the specimen under study, can then be plotted out directly as in any normal form of spectroscopy. A good example of the kind of information that can be obtained by applying Mössbauer techniques to biochemical compounds is afforded by the work of Lang and Marshall[22] on haemoglobin. They studied a large number of different derivatives of the haemoglobin, all of which had been artificially enriched in Fe^{57} by feeding this to rats. The results they obtained could be divided into two main groups, corresponding to ferrous and ferric derivatives. The only splittings they observed in the former cases were due to a nuclear quadrupole interaction, as shown in Fig. 7.9(*a*). This shows the simple doublet that is observed from reduced haemoglobin; the splitting arises from the interaction of the quadrupole moment of the Fe^{57} nucleus with the electrostatic field of its molecular surroundings.

When the Mössbauer spectra of ferric haemoglobin derivatives are plotted, however, further splittings are observed as shown in Fig. 7.9(*b*). These spectra are from haemoglobin fluoride, which, it will be remembered, is in the high spin state with $S = \frac{5}{2}$ and has a large zero field splitting between the component spin levels. These additional splittings in the Mössbauer spectrum can, in fact, be used to estimate the zero-field splitting parameter D, they can also be checked by taking further Mössbauer spectra in the presence of an applied external magnetic field.

Such spectra are shown in Fig. 7.9(*c*), and a detailed analysis of these[22] gives an estimate for D of about 7 cm^{-1}, which may be compared with that obtained for E.S.R. Lang and Marshall[22] carried out a large number of such analyses on the various derivatives which they studied and made direct comparisons between their own deductions on the energy level states of the iron atom with those obtained earlier from the electron resonance measurements.

Since this book is not really concerned with Mössbauer spectroscopy as such, the way in which these spectra can be interpreted in detail will not be pursued. Sufficient description may have been given, however, to show that the determination of these fine energy-level splittings, from the molecular interactions around the iron atom, can be directly correlated with the information obtained from the electron resonance measurements. These two forms of spectroscopy can thus be very helpful complementary techniques in the study of any biological molecules containing iron atoms.

The real point of this last section, however, has been to illustrate the interdependence of the various techniques, which are now being used to investigate biochemical and biological compounds in detail. The whole field of molecular biology, and the growing interest in the life sciences as a whole, has depended and will continue to depend on the precision previously associated with physical and chemical determinations, but now being applied and made available in the biological sphere. It is evident that high precision is the common feature possessed by all the new techniques coming into this growing point of science. Thus, it has been the precise measurements of intensity and angle of X-ray crystallographers, and their ability to handle and analyse these directly by computer techniques, that has enabled them to elucidate the three-dimensional structure of the very complex biochemical molecules. In a similar way, the very fine precision of energy-level measurement that the Mössbauer effect allows, has made detailed calculations of nuclear-molecular interaction possible inside the biochemical molecules themselves. In the same way, the high precision in microwave-frequency determination and magnetic-field measurement, has allowed electron resonance to differentiate precisely between the g-values and hyperfine splittings associated with free radicals, or metal ions, in the biochemical or biological molecules. It has been this ability to determine quite specifically the presence of unpaired electrons, and hence of transient radicals or other species, and also to characterize them precisely, that has given electron resonance such potentiality in these fields of biochemical research.

It could be said in closing that the application of electron resonance in these fields is a very good example of the new way in which

U*

science itself is moving. It is a technique invented not very long ago by the physicists, initially employed purely in physical measurements, and then taken over by the chemists. Most of the detailed theory of molecular orbital methods, and ligand fields, has indeed come from the theoretical chemists, and now both the techniques and theories, developed in the other two sciences, are moving rapidly into the field of molecular biology. In common with many other new ideas and methods of investigation, it is thus moving from one area of science to another, and in the process, it is playing its part in helping to integrate them into one unified whole.

References

1. Bray, R. C., *The Enzymes*, p. 533 (Academic Press, New York, 1963).
2. Rajagopalan, K. V., Fridovich, I., and Handler, P., *J. Biol. Chem.*, 1962, **237**, 922, and 1964, **239**, 2022.
3. Beinert, H., and Palmer, G., *Advances in Enzymology*, 1965, **27**, 105.
4. Beinert, H., and Orme-Johnson, W. H., *Magnetic Resonance in Biological Systems*, p. 237 (Pergamon Press, London, 1967).
5. Beinert, H., and Hemmerich, P., *Biochem. Biophys. Res. Comm.*, 1965, **18**, 212.
6. Hemmerich, P., *Helv. Chim. Acta.*, 1964, **47**, 464.
7. Beinert, H., and Orme-Johnson, W. H., *op. cit.*, p. 234.
8. Deal, R. M., Ingram, D. J. E., and Srinivasan, R., *Electronic Magnetic Resonance and Solid Dielectrics*, p. 239 (North Holland Publishing Co., 1963).
9. Eisenberger, P., and Pershan, P. S., *J. Chem. Phys.*, 1967, **47**, 3327.
10. Hyde, J. S., and Maki, A. H., *J. Chem. Phys.*, 1964, **40**, 3117.
11. Hyde, J. S., *Magnetic Resonance in Biological Systems*, p. 63 (Pergamon Press, 1967).
12. Hyde, J. S., *J. Chem. Phys.*, 1965, **43**, 1806.
13. Ranon, U., and Hyde, J. S., *Phys. Rev.*, 1966, **141**, 259.
14. Muller, F., Ehrenberg, A., and Erikson, L. E. G., *Magnetic Resonance in Biological Systems*, p. 281 (Pergamon Press, 1967).
15. Eisinger, J., Schulman, R. G., and Syymanski, B. M., *J. Chem. Phys.*, 1962, **36**, 1721.
16. Overhauser, A. W., *Phys. Rev.*, 1953, **92**, 411.
17. Ohnishi, S., and McConnell, H. M., *J. Am. Chem. Soc.*, 1965, **87**, 2293.
18. Griffith, O. H., and McConnell, H. M., *Proc. Nat. Acad. Sci.*, 1966, **55**, 8, and 1966, **55**, 708, and 1968, **61**, 12.
19. Griffith, O. H., Cornell, D. W., and McConnell, H. M., *J. Chem. Phys.*, 1965, **43**, 2909.
20. McConnell, H. M., *Magnetic Resonance in Biological Systems*, p. 319 (Pergamon Press, 1967).
21. Perutz, M. F., Steinrauf, L. K., Stockell, A., and Bangham, A. D., *J. Mol. Biol.*, 1959, **1**, 402.
22. Lang, G., and Marshall, W., *Proc. Phys. Soc.*, 1966, **87**, 3.

23. Gonser, U., and Grant, R. W., *Applied Phys. Letters*, 1963, **3**, 189, and *Science*, 1964, **143**, 680.
24. Tripathi, K. C., *Magnetic Resonance in Biological Systems*, p. 403 (Pergamon Press, 1967).
25. Maeda, Y., Higashimura, T., and Morita, Y., *Biochem. and Biophys. Res. Comm.*, 1967, **29**, 362, 680.
26. Maeda, Y., *J. Phys. Soc. Japan*, 1968, **24**, 151.
27. Bearden, A. J., and Moss, T. H., *Magnetic Resonance in Biological Systems*, p. 391 (Pergamon Press, 1967).
28. Johnson, C. E., Bray, R. C., and Knowles, P. F., *Magnetic Resonance in Biological Systems*, p. 417 (Pergamon Press, 1967).
29. Ehrenberg, A., Eriksson, L. E. G., and Hyde, J. S., *Biochem. Biophys. Acta*, 1968, **167**, 482.
30. Rajagopalan, K. V., Handler, P., Palmer, G., and Beinert, H., *J. Biol. Chem.*, 1968, **243**, 3784.

AUTHOR INDEX

SUBJECT INDEX